Operators vs Quantifiers

In this volume, seven experts in logic and semantics examine reasons for using the intensional operator approach over the variable binding approach and vice versa.

In logic and semantics there are two alternative tools that can be applied to many types of embedding phrases (modal, temporal, etc.): the intensional operator approach and the variable binding approach. A rivalry between operators and quantifiers occurs in many areas of semantics: e.g., tense, modality, locational operators, and epistemic modality. There are areas where the operator approach dominates and areas where quantifiers prevail. Sometimes, as in the case of tense, roles have switched, and where one approach used to dominate, the other has taken over. This volume features contributions from leading experts in the field that examine the following questions:

- What exactly is at stake in a choice between the alternatives?
- Are there any principled reasons for deciding which approach to use in which case?
- Should we be purists and eliminate one approach completely in favour of the other?

Operators vs Quantifiers will be a key resource for academics, researchers, and advanced students of philosophy, linguistics, computer science, and mathematics. This book was originally published as a special issue of *Inquiry*.

Max Kölbel is Professor of Philosophy at the University of Vienna, Austria.

David Rey is Assistant Professor at the Department of Philosophy of Universidad del Valle, Colombia.

Operators vs Quantifiers

Edited by
Max Kölbel and David Rey

LONDON AND NEW YORK

First published 2024
by Routledge
4 Park Square, Milton Park, Abingdon, Oxon, OX14 4RN

and by Routledge
605 Third Avenue, New York, NY 10158

Routledge is an imprint of the Taylor & Francis Group, an informa business

Introduction © 2024 Max Kölbel and David Rey
Chapters 1–6 © 2024 Taylor & Francis

All rights reserved. No part of this book may be reprinted or reproduced or utilised in any form or by any electronic, mechanical, or other means, now known or hereafter invented, including photocopying and recording, or in any information storage or retrieval system, without permission in writing from the publishers.

Trademark notice: Product or corporate names may be trademarks or registered trademarks, and are used only for identification and explanation without intent to infringe.

British Library Cataloguing-in-Publication Data
A catalogue record for this book is available from the British Library

ISBN13: 978-1-032-58095-1 (hbk)
ISBN13: 978-1-032-58096-8 (pbk)
ISBN13: 978-1-003-44254-7 (ebk)

DOI: 10.4324/9781003442547

Typeset in Myriad Pro
by codeMantra

Publisher's Note
The publisher accepts responsibility for any inconsistencies that may have arisen during the conversion of this book from journal articles to book chapters, namely the inclusion of journal terminology.

Disclaimer
Every effort has been made to contact copyright holders for their permission to reprint material in this book. The publishers would be grateful to hear from any copyright holder who is not here acknowledged and will undertake to rectify any errors or omissions in future editions of this book.

Contents

	Citation Information	vi
	Notes on Contributors	vii
	Introduction *Max Kölbel and David Rey*	1
1	Are quantifiers intensional operators? *Kai F. Wehmeier*	5
2	Binding bound variables in epistemic contexts *Brian Rabern*	27
3	Operators vs. quantifiers: the view from linguistics *Ariel Cohen*	58
4	Confessions of a schmentencite: towards an explicit semantics *Jonathan Schaffer*	87
5	Positing covert variables and the quantifier theory of tense *Matthew McKeever*	118
6	Looking backwards in type logic *Jan Köpping and Thomas Ede Zimmermann*	140
	Index	167

Citation Information

The following chapters were originally published in the journal *Inquiry*, volume 64, issue 5-6 (2021). When citing this material, please use the original page numbering for each article, as follows:

Chapter 1
Are quantifiers intensional operators?
Kai F. Wehmeier
Inquiry, volume 64, issue 5-6 (2021), pp. 511-532

Chapter 2
Binding bound variables in epistemic contexts
Brian Rabern
Inquiry, volume 64, issue 5-6 (2021), pp. 533-563

Chapter 3
Operators vs. quantifiers: the view from linguistics
Ariel Cohen
Inquiry, volume 64, issue 5-6 (2021), pp. 564-592

Chapter 4
Confessions of a schmentencite: towards an explicit semantics
Jonathan Schaffer
Inquiry, volume 64, issue 5-6 (2021), pp. 593-623

Chapter 5
Positing covert variables and the quantifier theory of tense
Matthew McKeever
Inquiry, volume 64, issue 5-6 (2021), pp. 624-645

Chapter 6
Looking backwards in type logic
Jan Köpping and Thomas Ede Zimmermann
Inquiry, volume 64, issue 5-6 (2021), pp. 646-672

For any permission-related enquiries please visit:
http://www.tandfonline.com/page/help/permissions

Notes on Contributors

Ariel Cohen, Department of Foreign Literatures and Linguistics, Ben-Gurion University of the Negev, Beer Sheva, Israel.

Max Kölbel, Department of Philosophy, University of Vienna, Austria.

Jan Köpping, Linguistics, Goethe-University Frankfurt am Main, Germany.

Matthew McKeever, ConceptLab, University of Oslo, Norway.

Brian Rabern, School of Philosophy, Psychology, and Language Sciences, University of Edinburgh, UK.

David Rey, Department of Philosophy, Universidad del Valle, Colombia.

Jonathan Schaffer, Department of Philosophy, Rutgers University, New Brunswick, USA.

Kai F. Wehmeier, Center for the Advancement of Logic, its Philosophy, History, and Applications, University of California, Irvine, USA.

Thomas Ede Zimmermann, Linguistics, Goethe-University Frankfurt am Main, Germany.

Introduction

Max Kölbel and David Rey

A number of age-old and recent debates in several branches of philosophy depend in some way on the competition between two apparently different approaches to modelling linguistic phenomena: on the one hand the approach that employs binding expressions (e.g. quantifiers) and variables and on the other hand the approach that employs intensional operators.

For example, the most standard approach to modality treats expressions like "possibly" as intensional operators. A competing approach treats them as binding quantifiers ("There is a possibility *x*, such that ..."). In the treatment of tense, the situation is reverse: the traditional intensional treatment of phenomena of tense, as introduced by Prior, is now a minority view, while a quantificational/referential treatment has become standard. Analogous alternatives present themselves in the treatment of countless other embedding expressions that shift some features: "somewhere", "in some way", "on every standard", "given what he knows", etc.

In fact, even "some" and "all" can be modeled not only as variable-binding expressions but also as operators, as in an Aristotelian syllogistic or term logic. Famously, the inventor of logic himself started logic off without variables, which prompted Geach (1968) to complain that Aristotle had started it off on the wrong foot. An unjustified complaint, as has been shown by many, including Schönfinkel (1924), Quine (1960), and Sommers (1982). The present-day controversy in semantics between the standard approach and a variable-free semantics à la Jacobson (1999) seems to be the contemporary version of this dispute.

It may be a historical accident that the most mainstream semantic approaches combine an operator treatment of some phenomena with a quantifier treatment of other phenomena. Only some purists argue for a variable-free pure operator approach—e.g. Jacobson—or for an operator-free pure variable binding approach—e.g. Schaffer (see Chapter 4 of this volume).

It therefore seems worth exploring what exactly is at stake in a choice between the alternatives: Do we need to employ both operators *and* quantifiers in a mixed semantics? If so, are there any principled reasons for deciding which approach to use in which case? Are there reasons of theoretical convenience or elegance? If we do not need to employ both, which of the two purist approaches should we choose? This volume collects a group of contributions that try to make progress with these questions.

This volume contains six chapters written by seven experts from the fields of philosophy, logic, and linguistics. The titles and authors of the chapters are listed below:

1. Are quantifiers intensional operators?
   Kai F. Wehmeier

2. Binding bound variables in epistemic contexts
   Brian Rabern

3. Operators vs. quantifiers: the view from linguistics
   Ariel Cohen

4. Confessions of a schmentencite: towards an explicit semantics
   Jonathan Schaffer

5. Positing covert variables and the quantifier theory of tense
   Matthew McKeever

6. Looking backwards in type logic
   Jan Köpping and Thomas Ede Zimmermann

Chapters 1 and 6 deal with type-theoretical formal languages. In Chapter 1, Wehmeier examines different formulations of type theory for languages of higher order predicate logic. He argues that the most satisfactory type-theoretical treatment of quantification is obtained by adopting a Fregean approach in which variable-binders—quantifiers and lambda abstractors—do not count as intensional operators. This result, according to Wehmeier, suggests that we should give a negative answer to the question posed in the title of Chapter 1. In Chapter 2, Köpping and Zimmermann take a look at two intensional strategies to overcome the expressive limitations of first-order modal languages. One strategy is to switch to a higher order system along the lines of Montague's intensional logic. The other is to add backwards-looking operators to a given modal language.[1] Köpping

---

[1] For discussion of this strategy, see Saarinen 1979 and Yanovich 2015. Although Saarinen and Yanovich seem to share the same intuitive notion of a backwards-looking operator, Saarinen formalizes backwards-looking operators in game-theoretical terms, whereas Yanovich adopts a multiple-index framework. Yanovich's backwards-looking operators are similar to the operators K_n and R_n of Vlach (1973, Appendix) and the operators ***actually**$_n$* and *Ref$_n$* of Cresswell (1990).

and Zimmermann characterize a (two-dimensional) language of higher order intensional logic and argue that the addition of backwards-looking operators does not increase its expressive power, which amounts to saying that the second strategy becomes redundant once the first strategy is adopted.

Chapters 2–5 focus on the semantics of natural languages. In Chapter 2, Rabern points out that, contrary to folk wisdom, a quantifier does not automatically become vacuous when it is prefixed to a formula with bound variables. Quantifiers and operators, he argues, can bind bound variables.[2] According to Rabern, this observation reveals that quantifying into epistemic constructions is no obstacle to assignment-shifting treatments of attitude verbs and epistemic modals. The proper analysis of modality, Rabern suggests, may be one that applies Kripkean semantics to metaphysical modality and Lewisian counterpart semantics to epistemic modality. In Chapter 3, Cohen explores the explanatory merits of the quantifier approach and the operator approach from a cross-linguistic perspective. By considering what linguists have found about a number of languages from different families, Cohen proposes a generalization which connects the availability of indexicals of a given kind in a language with the information that is represented in that language at the INFL level. He argues that the quantifier approach accounts for this generalization, whereas the operator approach fails to do so. In Chapter 4, Schaffer defends a view that treats individuals, worlds, and times in a unified extensional manner. This view, which he calls *explicit semantics*, posits explicit world, time, and individual variables in the grammatical structure of natural-language sentences and conceives propositions as being true or false simpliciter. Schaffer motivates explicit semantics by examining certain parallels between the temporalism-eternalism debate in the philosophy of time and the contingentism-necessitarianism debate in the philosophy of modality. He also compares explicit semantics with its main theoretical rivals and concludes that it is superior to them. Finally, Chapter 5 discusses the status of covert temporal variables in formal semantics. McKeever considers one important objection to the postulation of such variables, namely that their existence is not supported by syntactic evidence. He argues that, despite the absence of syntactic motivation, formal semanticists have good methodological reasons to postulate covert variables that refer to times and quantifiers that bind those variables.

---

[2] To understand Rabern's view properly, it is worth keeping in mind that his notion of binding is semantic, whereas his notion of bound variable is syntactic (see especially footnote 6 of Chapter 2).

Our brief summary of the chapters included in this volume reveals something worth noting: most of our contributors seem to adopt quantifier-oriented views. Those theorists who happen to prefer the quantificational approach may take this result as a natural consequence of the superiority of the extensional machinery of variables and variable-binders. For the advocates of the operator approach, on the other hand, the chapters of this volume provide a battery of empirical and theoretical challenges that must be faced by anyone who wants to vindicate the intensional paradigm.

References

Cresswell, M. J. 1990. *Entities and Indices*. Dordrecht: Kluwer Academic Publishers.

Geach, P. T. 1968. *A History of the Corruptions of Logic: An Inaugural Lecture*. Leeds, England: Leeds University Press.

Jacobson, P. 1999. "Towards a Variable-Free Semantics." *Linguistics and Philosophy* 22 (2): 117–184.

Quine, W. V. 1960. "Variables Explained Away." *Proceedings of the American Philosophical Society* 104 (3): 343–347.

Saarinen, E. 1979. "Backwards-Looking Operators in Tense Logic and in Natural Language." In *Game-Theoretical Semantics: Essays on Semantics by Hintikka, Carlson, Peacocke, Rantala, and Saarinen*, edited by E. Saarinen, 215–244. Dordrecht: D. Reidel Publishing Company.

Schönfinkel, M. 1924. "Über die Bausteine der mathematischen Logik." *Mathematische Annalen* 92: 305–316 (referred to by Quine in "Variables explained away").

Sommers, F. 1982. *The Logic of Natural Language*. Oxford: Oxford University Press.

Vlach, F. 1973. *'Now' and 'Then': A Formal Study in the Logic of Tense Anaphora*. Ph.D. Dissertation, UCLA.

Yanovich, I. 2015. "Expressive Power of "Now" and "Then" Operators." *Journal of Logic, Language and Information* 24 (1): 65–93.

Are quantifiers intensional operators?

Kai F. Wehmeier

ABSTRACT
In this paper, I ask whether quantifiers are intensional operators, with variable assignments playing the role of indices. Certain formulations of extensional type theory suggest an affirmative answer, but the most satisfactory among them suffer from a contamination of their semantic ontology with syntactic material. I lay out 'Fregean' versions of extensional type theory that are free from syntactic contamination and suggest a negative answer to our question.

1. Introduction

By an *intensional operator* (simpliciter) I shall mean any expression O that combines with sentences ϕ to form well-formed expressions (usually sentences) Oϕ and whose extension $[\![O]\!]^{\mathfrak{M},i}$ at an index i in a model $\mathfrak{M}$ takes sentential intensions, i.e. functions from indices to truth values, as arguments.[1] A *strongly intensional operator* is an intensional operator O such that there are models $\mathfrak{M}$, indices i and sentential intensions δ and ε for which $[\![O]\!]^{\mathfrak{M},i}(\delta) \neq [\![O]\!]^{\mathfrak{M},i}(\varepsilon)$ even though $\delta(i) = \varepsilon(i)$. A *weakly intensional operator* is an intensional operator that isn't strongly intensional. An *extensional operator* is a sentential operator O whose extension $[\![O]\!]^{\mathfrak{M},i}$ at any index i in any model $\mathfrak{M}$ takes truth values as arguments.

In the present paper, I examine the view that *quantifiers* are intensional operators, with variable assignments playing the role of indices.[2] Certain

---

[1]Thus categorial grammar places intensional operators in some category X/S, where X is usually S, and within intensional type theory they are of some type $\langle\langle s, t\rangle, a\rangle$. I here assume a framework for intensional semantics of the sort employed by von Fintel and Heim (2011), where, if O is of category S/S, the type a is t.

[2]I will be somewhat sloppy in my employment of the terms *quantifier* and *operator*, using them at times for expressions, at times for their extensions. I will also shift back and forth between taking the existential quantifier to be ∃ and taking it to be ∃v. More pedantic distinctions are no doubt desirable but would make this paper an even more soporific read.

formulations of extensional type theory indeed support such a view.[3] Others, however, do not, and anyway the view is at least *prima facie* in tension with Frege's (1893) insight that the first-order quantifiers are ordinary second-order functions, an insight central to present-day generalized quantifier theory.

In Section 2, we first examine what is perhaps the most familiar version of type theory, here called $\mathcal{L}_T$, where the quantifiers are introduced syncategorematically (i.e. don't have lexical entries). This formulation of the theory doesn't really help settle the question whether quantifiers are intensional operators: On the one hand, it suggests a negative answer, because it doesn't assign the quantifiers any extensions at all, so that our definition of intensional operators is inapplicable. On the other hand, it suggests a positive answer, because its syncategorematic quantifier clauses are structurally identical to the familiar syncategorematic clauses for modal and temporal operators. Since there are independent reasons to be dissatisfied with $\mathcal{L}_T$, we next turn to another popular version of extensional type theory, $\mathcal{L}_T^\lambda$, in which syncategorematic lambda abstraction enables lexical entries for the quantifiers. In $\mathcal{L}_T^\lambda$, the quantifiers themselves are not intensional operators. Lambda abstraction, however, has an intensional flavor, which the quantifier-lambda combination inherits, and it's not implausible that it is really this *combination* that corresponds to the syncategorematic quantifier of $\mathcal{L}_T$. So $\mathcal{L}_T^\lambda$ doesn't unequivocally answer our question either, and in any case, the lack of a lexical entry for abstraction in $\mathcal{L}_T^\lambda$ is still unsatisfactory.

We thus turn, in Section 3, to two formulations of type theory in which *every* operator, including the variable-binders, has a lexical entry. One version, $\mathcal{L}_{T^s}^\lambda$, includes the lambda operator; here the quantifiers themselves behave as they do in $\mathcal{L}_T^\lambda$ and thus aren't intensional operators; the lambda operator, however, is. The other version, $\mathcal{L}_{T^s}$, doesn't include lambda abstraction, and here the quantifiers themselves literally *are* intensional operators.

In Section 4, I argue that the semantics for $\mathcal{L}_{T^s}^\lambda$ and $\mathcal{L}_{T^s}$, though compositional and in accord with Frege's Conjecture, are ontologically unsatisfactory due to syntactic contamination of the type hierarchy.

Section 5 introduces a Frege-inspired treatment of extensional type theory that escapes this contamination issue. Crucially, it turns out that

---

[3]For a somewhat recent advocate of a form of this view, see Belnap (2005), who emphasizes the 'massive likeness of the practices of "extensional" and "intensional" semantics' (3) and goes so far as to claim that it is 'an illogical and harmful historical aberration that it is seldom made explicit that *quantifiers are non-truth-functional connectives*' (6; emphasis in the original).

under a Fregean construal, neither quantifiers nor abstraction operators are in any sense intensional operators.

In Section 6, I point out that Fregean type theories appear to be the most satisfactory options. As quantifiers aren't intensional operators in these best versions of type theory, I conclude that the appearance of quantifiers as intensional operators in other, inferior, systems cannot be taken as reflecting a fundamental semantic property. I also briefly raise the question whether, with respect to natural language, the concept of an intensional operator is a fruitful one.

2. Syncategorematic type theory

The set **T** of *extensional types* is inductively generated from the basic types e and t by allowing the formation of a type $\langle a, b \rangle$ from any types a and b. The *extensional type hierarchy* over a non-empty set D is the family $(D_a)_{a \in \mathbf{T}}$ such that

- $D_e = D$ (*individuals*);
- $D_t = \{0, 1\}$ (*truth values*);
- $D_{\langle a, b \rangle} = D_b^{D_a}$ (*functions from D_a to D_b*).

The type-theoretical language $\mathcal{L}_\mathbf{T}$ has as *primitive symbols*

- for each $a \in \mathbf{T}$, a countably infinite set $\mathcal{C}_a$ of ('non-logical') *constants* of type a;
- for each $a \in \mathbf{T}$, a countably infinite set V_a of *variables* of type a;
- the sentential *connectives* $\neg$ and $\wedge$;
- the existential *quantifier* $\exists$.

The *well-formed expressions* (wfe's) of $\mathcal{L}_\mathbf{T}$ and their types are defined as follows:

(1) Every $c \in \mathcal{C}_a$ is an $\mathcal{L}_\mathbf{T}$-wfe of type a.
(2) Every $v \in V_a$ is an $\mathcal{L}_\mathbf{T}$-wfe of type a.
(3) The connectives $\neg$ and $\wedge$ are $\mathcal{L}_\mathbf{T}$-wfe's of types $\langle t, t \rangle$ and $\langle t, \langle t, t \rangle \rangle$, respectively.
(4) If α is an $\mathcal{L}_\mathbf{T}$-wfe of type $\langle a, b \rangle$ and β an $\mathcal{L}_\mathbf{T}$-wfe of type a, then $\alpha\beta$ is an $\mathcal{L}_\mathbf{T}$-wfe of type b.
(5) If ϕ is an $\mathcal{L}_\mathbf{T}$-wfe of type t and v a variable of any type, $\exists v \phi$ is an $\mathcal{L}_\mathbf{T}$-wfe of type t.

A *model* $\mathfrak{M}$ for $\mathcal{L}_T$ is a pair $\langle D, \mathcal{I} \rangle$ consisting of a non-empty set D (the *domain* of $\mathfrak{M}$) and an *interpretation function* $\mathcal{I}$ that maps, for each $a \in \mathbf{T}$, every element of $\mathcal{C}_a$ to an element of D_a.

A *variable assignment over D* is a function that maps, for each $a \in \mathbf{T}$, every member of V_a to an element of D_a. A variable assignment *in a model* $\mathfrak{M}$ is a variable assignment over the model's domain. We let $\aleph_D$ be the set of all variable assignments over D.

We define the *extensions* $[\![\alpha]\!]^{\mathfrak{M},g}$ of $\mathcal{L}_T$-wfe's α relative to a model $\mathfrak{M} = \langle D, \mathcal{I} \rangle$ and a variable assignment g in such a way that whenever α is of type a, $[\![\alpha]\!]^{\mathfrak{M},g}$ is in D_a. Clauses 1–3 are lexical entries, while 4 and 5 are composition rules.[4]

(1) For $a \in \mathbf{T}$ and $\alpha \in \mathcal{C}_a$, $[\![\alpha]\!]^{\mathfrak{M},g} = \mathcal{I}(\alpha)$.
(2) For $a \in \mathbf{T}$ and $v \in V_a$, $[\![v]\!]^{\mathfrak{M},g} = g(v)$.
(3) $[\![\neg]\!]^{\mathfrak{M},g} = \underline{\lambda} n \in \{0,1\}.\ 1 - n$ and
    $[\![\wedge]\!]^{\mathfrak{M},g} = \underline{\lambda} n \in \{0,1\}.\ \underline{\lambda} m \in \{0,1\}.\ \min\{n,m\}$.
(4) 'Functional Application': $[\![\alpha\beta]\!]^{\mathfrak{M},g} = [\![\alpha]\!]^{\mathfrak{M},g}([\![\beta]\!]^{\mathfrak{M},g})$.
(5) $[\![\exists v\phi]\!]^{\mathfrak{M},g} = \max\{[\![\phi]\!]^{\mathfrak{M},h} \mid h \in \aleph_D, h \sim_v g\}$, where $\sim_v$ is the relation of v-variance between assignments.

Strictly speaking, asking whether the quantifier ∃v as it occurs in $\mathcal{L}_T$ is an intensional operator doesn't make sense, because our definition refers to the operator's extension at an index, and while it is perhaps natural to think of assignments as indices, our semantics for $\mathcal{L}_T$ doesn't assign ∃v an extension relative to an assignment. In a broader sense, however, it's hard to deny that ∃v does *look* like an intensional operator. For consider the syncategorematic clause for the possibility operator ◊ of modal logic that one finds in logic textbooks: ◊ϕ is true at world w if ϕ is true at some world v that is accessible from w. This is structurally identical to the syncategorematic quantifier clause of $\mathcal{L}_T$: ∃vϕ is true at assignment g if ϕ is true at some assignment h that is a v-variant of g. Still, it would be nice if we could confirm this impression in a system with a lexical entry for the quantifier – after all, in semantics textbooks the possibility operator *is* given a lexical entry, revealing it to take intensions as arguments.

Moreover, we should trust a formal system only to the extent that it exhibits theoretical virtues. One relevant virtue here is compositionality.

---

[4]Underlined lambdas, as they occur in the clauses for ¬ and ∧, belong to the semantic metalanguage. We will shortly encounter object-linguistic lambdas as well, which will not be adorned by underlining.

It's clear that extensions relative to a fixed $\mathfrak{M}$ and g do *not* abide by the principle of compositionality: by clause 5, it is not just the extension of ϕ relative to $\mathfrak{M}$ and g that figures into the determination of the extension of $\exists v \phi$ relative to $\mathfrak{M}$ and g; rather, we must take into account the extensions of ϕ relative to $\mathfrak{M}$ and h for a whole range of assignments h. So let the *assignment intension* $[\![\alpha]\!]^{\mathfrak{M}}$ of α in $\mathfrak{M}$ be the function with domain $\aleph_D$ that maps each assignment g over D to the extension $[\![\alpha]\!]^{\mathfrak{M},g}$ of α relative to $\mathfrak{M}$ and g. Assignment intensions have a better chance of satisfying compositionality, since $[\![\exists v \phi]\!]^{\mathfrak{M}}$ is the function that maps any g to $\max\{[\![\phi]\!]^{\mathfrak{M}}(h) \mid h \in \aleph_D, h \sim_v g\}$ and is therefore a function of the assignment intension $[\![\phi]\!]^{\mathfrak{M}}$ of ϕ in $\mathfrak{M}$.

Even so, it is debatable whether, strictly speaking, our assignment intensions for $\mathcal{L}_T$ are compositional. The principle of compositionality requires that for every syntactic rule ν there is a semantic operation r_ν such that the meaning $\mu(\nu(u_1, \ldots, u_n))$ of $\nu(u_1, \ldots, u_n)$ is equal to $r_\nu(\mu(u_1), \ldots, \mu(u_n))$, where $\mu(u)$ is the meaning of the expression u.[5] Now our syntax suggests that negations and existential quantifications are obtained by the same syntactic composition rule, to wit, concatenation: negations $\neg \phi$ arise by concatenating $\neg$ and ϕ; existential quantifications $\exists v \phi$ arise by concatenating $\exists v$ and ϕ. By compositionality, they should be subject to the same semantic rule; but that is not so – negations are evaluated according to the functional application rule 4, while existential quantifications are evaluated according to the syncategorematic existential quantifier rule 5.[6]

Besides a possible violation of Compositionality, there are a couple of additional reasons to find the quantifier clause unsatisfactory. One such reason is that the syncategorematic introduction of the existential quantifier, that is, the inclusion of a special-purpose rule for wfe's governed by this particular operator, obscures the semantic parallelism between the various kinds of quantifiers: 'some', 'all', 'no', 'the', 'exactly one', 'at most seventeen', 'infinitely many' etc. In a language that includes several of these, a tailor-made syncategorematic semantic rule would be required in each case, which fails to bring out the fact, discovered by Frege and central to contemporary generalized quantifier theory, that all of these quantifiers can be construed as higher order functions. In other words, the semantic composition rule is intuitively the same in all these cases

---

[5] See e.g. Pagin and Westerståhl (2010).
[6] Technically there's some wiggle room: we could make the syntactic rules for negation and existential quantification artificially distinct, e.g. by introducing parentheses and writing the negation of ϕ as $\neg(\phi)$ but the existential quantification of ϕ as $(\exists v \phi)$; this difference would allow us to pair them with different semantic rules. This smells of cheating, though.

(viz., functional application), the differences arising out of the *lexical* meanings of the quantifier expressions; this isn't reflected in a semantics with no lexical entries, but a whole smorgasbord of syncategorematic rules, for the quantifiers.

A final reason for dissatisfaction with $\mathcal{L}_T$ is specific to a particular way of pursuing compositional semantics, espoused, for example, by Heim and Kratzer (1998) and guided by 'Frege's Conjecture', according to which semantic composition is functional application.[7] Rule 5 for existentially quantified wfe's isn't an instance of functional application and therefore violates Frege's Conjecture; so anyone guided by Frege's Conjecture ought to be wary about syncategorematic clauses.

There is thus considerable pressure to replace the syncategorematic treatment of the existential quantifier with a categorematic one. One way to attempt this would be to count strings ∃v as wfe's, assign them a type, let rule 5 in the definition of wfe's be subsumed by clause 4, directly assign extensions $[\![\exists v]\!]^{\mathfrak{M},g}$ to the newly recognized lexical items ∃v and let the syncategorematic clause 5 in the definition of extensions be subsumed by the functional application clause 4, all the while ensuring that all wfe's of $\mathcal{L}_T$ retain their original extensions. As is well known, this is impossible.[8]

A more promising, and quite popular, strategy is to pass the syncategorematic buck from quantification to abstraction. That is, we can give the existential quantifier a *categorematic* treatment as long as we allow ourselves the *syncategorematic* introduction of an abstraction operator.[9] Concretely, we replace $\mathcal{L}_T$ with a new language $\mathcal{L}_T^\lambda$ whose primitive symbols are those of $\mathcal{L}_T$, except that the symbol ∃ is replaced with infinitely many symbols $\exists^a$, one for each $a \in \mathbf{T}$, and that the abstractor symbol λ as well as parentheses are added. The $\mathcal{L}_T^\lambda$-wfe's and their types are defined as follows.

(1) Every $c \in \mathcal{C}_a$ is an $\mathcal{L}_T^\lambda$-wfe of type a.
(2) Every $v \in V_a$ is an $\mathcal{L}_T^\lambda$-wfe of type a.

---

[7] Heim and Kratzer themselves violate Frege's Conjecture with their Predicate Modification and Predicate Abstraction rules. As they explain (1998, §4.3.2), Predicate Modification can be brought into the fold of functional application in a number of ways, and in any case, the rule is of no particular *logical* interest. Predicate Abstraction is another matter, to which we return below.
[8] If $\exists v \phi$ is a wfe of $\mathcal{L}_T$, ϕ must be of type t, and $\exists v \phi$ itself is also of type t. Hence ∃v would have to be of type $\langle t, t \rangle$. So for each $\mathfrak{M}$ and g, $[\![\exists v]\!]^{\mathfrak{M},g}$ would have to be a unary truth function. That's easily seen to be impossible. Cf. Gamut (1991, 101–102).
[9] Indeed this seems to be the purpose of Heim and Kratzer's Predicate Abstraction rule.

(3) The connectives $\neg$ and $\wedge$ are $\mathcal{L}_T^\lambda$-wfe's of types $\langle t,t \rangle$ and $\langle t, \langle t,t \rangle \rangle$, respectively.
(4) For any $a \in \mathbf{T}$, $\exists^a$ is an $\mathcal{L}_T^\lambda$-wfe of type $\langle \langle a,t \rangle, t \rangle$.
(5) If α is an $\mathcal{L}_T^\lambda$-wfe of type $\langle a,b \rangle$ and β an $\mathcal{L}_T^\lambda$-wfe of type a, then $(\alpha\beta)$ is an $\mathcal{L}_T^\lambda$-wfe of type b.
(6) If α is an $\mathcal{L}_T^\lambda$-wfe of type a and v a variable of type b, then $(\lambda v \alpha)$ is an $\mathcal{L}_T^\lambda$-wfe of type $\langle b,a \rangle$.

The λ-enriched language $\mathcal{L}_T^\lambda$ differs somewhat from the original language $\mathcal{L}_T$. For example, $\mathcal{L}_T^\lambda$, but not $\mathcal{L}_T$, contains quantified formulas in which no bound variable is present (such as $\exists^e P$, where P is of type $\langle e,t \rangle$). Moreover, the quantified formulas of $\mathcal{L}_T$ get a new look in $\mathcal{L}_T^\lambda$: Instead of the $\mathcal{L}_T$-formula $\exists v Pv$, the language $\mathcal{L}_T^\lambda$ has $(\exists^e(\lambda v(Pv)))$. Finally, $\mathcal{L}_T^\lambda$, but not $\mathcal{L}_T$, contains lambda-abstracts $(\lambda v \alpha)$ as wfe's. It's intuitively clear, however, that $\mathcal{L}_T$ is essentially a sublanguage of $\mathcal{L}_T^\lambda$. More precisely, let ι be the function from the $\mathcal{L}_T$-wfe's to the $\mathcal{L}_T^\lambda$-wfe's that is defined as follows.

(1) For $a \in \mathbf{T}$ and $\alpha \in \mathcal{C}_a$, $\iota(\alpha) = \alpha$.
(2) For $a \in \mathbf{T}$ and $v \in V_a$, $\iota(v) = v$.
(3) $\iota(\neg) = \neg$ and $\iota(\wedge) = \wedge$.
(4) $\iota(\alpha\beta) = (\iota(\alpha)\iota(\beta))$.
(5) $\iota(\exists v \phi) = (\exists^a(\lambda v \iota(\phi)))$, where v is of type a.

The function ι is clearly 1–1 and preserves types. It thus embeds $\mathcal{L}_T$ in $\mathcal{L}_T^\lambda$.

We define the extensions $[\![\alpha]\!]_\lambda^{\mathfrak{M},g}$ of $\mathcal{L}_T^\lambda$-wfe's α relative to $\mathfrak{M}$ and g as follows. Clauses 1–4 are lexical entries, clauses 5 and 6 composition rules.

(1) For $a \in \mathbf{T}$ and $\alpha \in \mathcal{C}_a$, $[\![\alpha]\!]_\lambda^{\mathfrak{M},g} = \mathcal{I}(\alpha)$.
(2) For $a \in \mathbf{T}$ and $v \in V_a$, $[\![v]\!]_\lambda^{\mathfrak{M},g} = g(v)$.
(3) $[\![\neg]\!]_\lambda^{\mathfrak{M},g} = \lambda n \in \{0,1\}.\ 1-n$ and
$[\![\wedge]\!]_\lambda^{\mathfrak{M},g} = \lambda n \in \{0,1\}.\ \lambda m \in \{0,1\}.\ \min\{n,m\}$.
(4) For $a \in \mathbf{T}$, $[\![\exists^a]\!]_\lambda^{\mathfrak{M},g} = \lambda f \in D_t^{D_a}.\ \max\{f(u) \mid u \in D_a\}$.
(5) (FA) For α of type $\langle a,b \rangle$ and β of type a, $[\![(\alpha\beta)]\!]_\lambda^{\mathfrak{M},g} = [\![\alpha]\!]_\lambda^{\mathfrak{M},g}([\![\beta]\!]_\lambda^{\mathfrak{M},g})$.
(6) If α is an $\mathcal{L}_T^\lambda$-wfe of type a and $v \in V_b$, $[\![(\lambda v \alpha)]\!]_\lambda^{\mathfrak{M},g} = \lambda u \in D_b.\ [\![\alpha]\!]_\lambda^{\mathfrak{M},g[v:=u]}$, where $g[v:=u]$ is the v-variant of g that maps v to u.

It's easy to see that ι is an *extensionally faithful* embedding of $\mathcal{L}_T$ in $\mathcal{L}_T^\lambda$ in the following sense. For every model $\mathfrak{M}$, variable assignment g over $\mathfrak{M}$, and $\mathcal{L}_T$-wfe α:

$$[\![\alpha]\!]^{\mathfrak{M},g} = [\![\iota(\alpha)]\!]_\lambda^{\mathfrak{M},g}.$$

The quantifiers have now been given a categorematic treatment. A quantifier over objects of type a receives type $\langle\langle a,t\rangle, t\rangle$; in particular, the quantifier $\exists^e$ is of type $\langle\langle e,t\rangle, t\rangle$, which accords well not only with Frege's view but also with contemporary generalized quantifier theory. Thus *prima facie* the existential quantifiers $\exists^a$ as they occur in $\mathcal{L}_T^\lambda$ do not look like intensional operators, because their arguments aren't intensions.

This does not really settle our question, however, for one might argue that the role played by ∃v in $\mathcal{L}_T$ is now played by the *combination* of the quantifier $\exists^a$ with the lambda operator λv, so in order to elucidate the logical status of ∃v in $\mathcal{L}_T$ we should look not at $\exists^a$ in isolation, but at $\exists^a \circ \lambda$v, as it were. If we combine the semantic clauses for $\exists^a$ and λv, however, we obtain a syncategorematic clause identical to the one that ∃v had in $\mathcal{L}_T$.

Indeed it's not clear that we've made substantive progress at all, for the categorematic treatment of the quantifier has come at the price of a *syncategorematic* introduction of the abstraction operator, and of the three objections to the semantics for $\mathcal{L}_T$, at least two therefore remain: the concern about compositionality *stricto sensu*, arising from the different semantic treatment of concatenation in the cases of negation and lambda abstraction, and the violation of Frege's Conjecture. So it would help if we could give λ a categorematic treatment consistent with Frege's Conjecture. Is that possible?

It depends. Within our current type-theoretical framework, we cannot in general produce an assignment intension for λv that allows evaluation of λ-terms by way of functional application.[10]

As we will see in the next section, it *is* possible, modulo substantial revisions to the framework, to give λ a categorematic treatment in accord with

---

[10] Let $\mathfrak{M}_0$ be a model whose domain is the set of natural numbers $\mathbb{N}$, in which the interpretation of the constant P (of type $\langle e,t\rangle$) maps every number to 0, and in which the interpretation of the constant Q (of type $\langle e,t\rangle$) maps every positive number to 0 but maps 0 to 1. Let v be a variable of type e and let g assign 1 to v. Obviously $[\![(\lambda v(Pv))]\!]_\lambda^{\mathfrak{M}_0}(g)$ is the constant function on $\mathbb{N}$ with value 0, and $[\![(\lambda v(Qv))]\!]_\lambda^{\mathfrak{M}_0}(g)$ is the function on $\mathbb{N}$ that maps 0 to 1 but every positive number to 0. Since $(\lambda v(Pv))$ is of type $\langle e,t\rangle$ and (Pv) is of type t, λv would have to be of type $\langle t, \langle e,t\rangle\rangle$. Then $[\![\lambda v]\!]_\lambda^{\mathfrak{M}_0}(g)$ would have to be a function from D_t to $D_t^{D_e}$, so that $[\![\lambda v]\!]_\lambda^{\mathfrak{M}_0}(g)([\![(Pv)]\!]_\lambda^{\mathfrak{M}_0}(g))$ would have to equal $[\![\lambda v]\!]_\lambda^{\mathfrak{M}_0}(g)([\![(Qv)]\!]_\lambda^{\mathfrak{M}_0}(g))$, given that $[\![(Pv)]\!]_\lambda^{\mathfrak{M}_0}(g) = 0 = [\![(Qv)]\!]_\lambda^{\mathfrak{M}_0}(g)$. But then we can't in general have $[\![(\lambda v\alpha)]\!]_\lambda^{\mathfrak{M}_0}(g) = [\![\lambda v]\!]_\lambda^{\mathfrak{M}_0}(g)([\![\alpha]\!]_\lambda^{\mathfrak{M}_0}(g))$.

Frege's Conjecture. Categorematicity, however, comes at a price, which we'll assess in due course.

3. Assignatory type theory

As mentioned, making the abstraction operator categorematic requires some changes to the type-theoretical framework.[11]

We begin by expanding the set of types. Let the set $\mathbf{T}^s$ of *assignatory types* (s-types for short) be defined as follows: e and t are s-types, and whenever a and b are s-types, so are $\langle a, b \rangle$ and $\langle s, a \rangle$. The *assignatory type hierarchy* (s-type hierarchy for short) over a non-empty domain D is the family $(D_a)_{a \in \mathbf{T}^s}$ such that

- $D_e = D$;
- $D_t = \{0, 1\}$;
- $D_{\langle a, b \rangle} = D_b^{D_a}$, provided that $a \neq s$;
- $D_{\langle s, a \rangle} = D_a^{\aleph_D}$.

The primitive symbols of the assignatory language $\mathcal{L}_{\mathbf{T}^s}^\lambda$ are those of $\mathcal{L}_{\mathbf{T}}^\lambda$, except that the symbol λ is replaced by infinitely many symbols λ_b^a, one for each pair $\langle a, b \rangle \in \mathbf{T} \times \mathbf{T}$. The $\mathcal{L}_{\mathbf{T}^s}^\lambda$-wfe's and their types are defined as follows.

(1) Every $c \in \mathcal{C}_a$ is an $\mathcal{L}_{\mathbf{T}^s}^\lambda$-wfe of type a.
(2) Every $v \in V_a$ is an $\mathcal{L}_{\mathbf{T}^s}^\lambda$-wfe of type a.
(3) The connectives $\neg$ and $\wedge$ are $\mathcal{L}_{\mathbf{T}^s}^\lambda$-wfe's of types $\langle t, t \rangle$ and $\langle t, \langle t, t \rangle \rangle$, respectively.
(4) For $a \in \mathbf{T}$, $\exists^a$ is an $\mathcal{L}_{\mathbf{T}^s}^\lambda$-wfe of type $\langle \langle a, t \rangle, t \rangle$.
(5) For $a, b \in \mathbf{T}$ and $v \in V_b$, $\lambda_b^a v$ is an $\mathcal{L}_{\mathbf{T}^s}^\lambda$-wfe of type $\langle \langle s, a \rangle, \langle b, a \rangle \rangle$.
(6) If α is an $\mathcal{L}_{\mathbf{T}^s}^\lambda$-wfe of type $\langle a, b \rangle$ and β an $\mathcal{L}_{\mathbf{T}^s}^\lambda$-wfe of type a, then $(\alpha\beta)$ is an $\mathcal{L}_{\mathbf{T}^s}^\lambda$-wfe of type b.
(7) If α is an $\mathcal{L}_{\mathbf{T}^s}^\lambda$-wfe of type $\langle \langle s, a \rangle, b \rangle$ and β an $\mathcal{L}_{\mathbf{T}^s}^\lambda$-wfe of type a, then $(\alpha\beta)$ is an $\mathcal{L}_{\mathbf{T}^s}^\lambda$-wfe of type b.

We define a function κ from the $\mathcal{L}_{\mathbf{T}}^\lambda$-wfe's to the $\mathcal{L}_{\mathbf{T}^s}^\lambda$-wfe's as follows.

(1) For $a \in \mathbf{T}$ and $\alpha \in \mathcal{C}_a$, $\kappa(\alpha) = \alpha$.
(2) For $a \in \mathbf{T}$ and $v \in V_a$, $\kappa(v) = v$.

---

[11] I believe this was first pointed out by Varzi (1993).

(3) $\kappa(\neg) = \neg$ and $\kappa(\wedge) = \wedge$.
(4) For $a \in \mathbf{T}$, $\kappa(\exists^a) = \exists^a$.
(5) If α is an $\mathcal{L}_\mathbf{T}^\lambda$-wfe of type $\langle a,b \rangle$ and β an $\mathcal{L}_\mathbf{T}^\lambda$-wfe of type a, $\kappa((\alpha\beta)) = (\kappa(\alpha)\kappa(\beta))$.
(6) If α is an $\mathcal{L}_\mathbf{T}^\lambda$-wfe of type a and v a variable of type b, $\kappa((\lambda \mathsf{v} \alpha)) = (\lambda_b^a \mathsf{v} \kappa(\alpha))$.

It's easy to see that κ is 1−1 and that $\kappa(\alpha)$ is of the same type as α. Hence κ embeds $\mathcal{L}_\mathbf{T}^\lambda$ in $\mathcal{L}_{\mathbf{T}^\mathfrak{s}}^\lambda$, and $\kappa \circ \iota$ embeds $\mathcal{L}_\mathbf{T}$ in $\mathcal{L}_{\mathbf{T}^\mathfrak{s}}^\lambda$.

A model $\mathfrak{M}$ for $\mathcal{L}_{\mathbf{T}^\mathfrak{s}}^\lambda$ is still a pair $\langle D, \mathcal{I} \rangle$, where $D \neq \emptyset$ and $\mathcal{I}$ is a function assigning to each member of $\mathcal{C}_a$ an element of D_a, for every $a \in \mathbf{T}$.[12]

Next we define the extensions $[\![\alpha]\!]_{\lambda,\mathfrak{s}}^{\mathfrak{M},g}$, relative to $\mathfrak{M}$ and g, of the well-formed $\mathcal{L}_{\mathbf{T}^\mathfrak{s}}^\lambda$-expressions α. As before, we let $[\![\alpha]\!]_{\lambda,\mathfrak{s}}^{\mathfrak{M}}$ be the function that maps each variable assignment g to $[\![\alpha]\!]_{\lambda,\mathfrak{s}}^{\mathfrak{M},g}$, that is, the *assignment intension* of α (relative to $\mathfrak{M}$). To reduce clutter, we often suppress the model superscript $\mathfrak{M}$; thus, for example, the assignment intension of α will be written $[\![\alpha]\!]_{\lambda,\mathfrak{s}}$. The clauses for extensions can then be given as follows. Clauses 1–5 are lexical entries, clauses 6 and 7 are composition rules.

(1) For $a \in \mathbf{T}$ and $\alpha \in \mathcal{C}_a$, $[\![\alpha]\!]_{\lambda,\mathfrak{s}}^g = \mathcal{I}(\alpha)$.
(2) For $a \in \mathbf{T}$ and $\mathsf{v} \in V_a$, $[\![\mathsf{v}]\!]_{\lambda,\mathfrak{s}}^g = g(\mathsf{v})$.
(3) $[\![\neg]\!]_{\lambda,\mathfrak{s}}^g = \lambda n \in \{0,1\}.\ 1 - n$ and $[\![\wedge]\!]_{\lambda,\mathfrak{s}}^g = \lambda n \in \{0,1\}.\ \lambda m \in \{0,1\}.\ \min\{n, m\}$.
(4) For $a \in \mathbf{T}$, $[\![\exists^a]\!]_{\lambda,\mathfrak{s}}^g = \lambda f \in D_{\langle a,t \rangle}.\ \max\{f(u) \mid u \in D_a\}$.
(5) For $a, b \in \mathbf{T}$ and $\mathsf{v} \in V_b$, $[\![\lambda_b^a \mathsf{v}]\!]_{\lambda,\mathfrak{s}}^g = \lambda f \in D_a^{\aleph_D}.\ \lambda r \in D_b.\ f(g[\mathsf{v} := r])$.[13]
(6) Extensional Functional Application: If α is an $\mathcal{L}_{\mathbf{T}^\mathfrak{s}}^\lambda$-wfe of type $\langle a, b \rangle$ and β an $\mathcal{L}_{\mathbf{T}^\mathfrak{s}}^\lambda$-wfe of type a, then $[\![(\alpha\beta)]\!]_{\lambda,\mathfrak{s}}^g$ is $[\![\alpha]\!]_{\lambda,\mathfrak{s}}^g([\![\beta]\!]_{\lambda,\mathfrak{s}}^g)$.
(7) Intensional Functional Application: If α is an $\mathcal{L}_{\mathbf{T}^\mathfrak{s}}^\lambda$-wfe of type $\langle \langle \mathfrak{s}, a \rangle, b \rangle$ and β an $\mathcal{L}_{\mathbf{T}^\mathfrak{s}}^\lambda$-wfe of type a, then $[\![(\alpha\beta)]\!]_{\lambda,\mathfrak{s}}^g$ is $[\![\alpha]\!]_{\lambda,\mathfrak{s}}^g([\![\beta]\!]_{\lambda,\mathfrak{s}})$, i.e. $[\![\alpha]\!]_{\lambda,\mathfrak{s}}^g(\lambda h \in \aleph_D.\ [\![\beta]\!]_{\lambda,\mathfrak{s}}^h)$.

Note that these extensions live in the *assignatory* type hierarchy over the model's domain. In particular, the extension $[\![\lambda_b^a \mathsf{v}]\!]_{\lambda,\mathfrak{s}}^g$ of $\lambda_b^a \mathsf{v}$ at g belongs to $D_{\langle \langle \mathfrak{s}, a \rangle, \langle b, a \rangle \rangle}$ and thus falls outside the extensional type hierarchy $(D_a)_{a \in \mathbf{T}}$ over D.

It can be shown that κ is an extensionally faithful embedding of $\mathcal{L}_\mathbf{T}^\lambda$ in $\mathcal{L}_{\mathbf{T}^\mathfrak{s}}^\lambda$ (so that $\kappa \circ \iota$ is an extensionally faithful embedding of $\mathcal{L}_\mathbf{T}$

---
[12]Since we have more types in $\mathbf{T}^\mathfrak{s}$ than in $\mathbf{T}$, we could allow non-empty sets of non-logical constants of these additional types. For the sake of simplicity, we don't.
[13]Such a clause for $\lambda \mathsf{v}$ was first proposed, I believe, by Rabern (2012, 399).

in $\mathcal{L}_{T^s}^\lambda$).[14] Moreover, the extensional faithfulness of ι and κ, and concomitantly of $\kappa \circ \iota$, immediately implies their *intensional faithfulness*, meaning that an expression and its translation always have the same assignment intension. Thus we've finally arrived at a language that faithfully embeds the original type-theoretical language $\mathcal{L}_T$ and has a fully categorematic semantics.

Before we consider the status of quantifiers as intensional operators in $\mathcal{L}_{T^s}^\lambda$, let us note that once we've swallowed the expansion of the type hierarchy to the assignatory type hierarchy and the presence of the intensional functional application rule, the λ-operator is no longer needed for a categorematic treatment of the quantifier: Let the primitive symbols of the *austere* assignatory language $\mathcal{L}_{T^s}$ be those of $\mathcal{L}_{T^s}^\lambda$ minus all symbols λ_b^a and the parentheses. The $\mathcal{L}_{T^s}$-wfe's and their types are defined as follows:

(1) Every $c \in C_a$ is an $\mathcal{L}_{T^s}$-wfe of type a.
(2) Every $v \in V_a$ is an $\mathcal{L}_{T^s}$-wfe of type a.
(3) The connectives $\neg$ and $\wedge$ are $\mathcal{L}_{T^s}$-wfe's of types $\langle t, t \rangle$ and $\langle t, \langle t, t \rangle \rangle$, respectively.
(4) For $a \in \mathbf{T}$ and $v \in V_a$, $\exists^a v$ is an $\mathcal{L}_{T^s}$-wfe of type $\langle \langle s, t \rangle, t \rangle$.
(5) If α is an $\mathcal{L}_{T^s}$-wfe of type $\langle a, b \rangle$ and β an $\mathcal{L}_{T^s}$-wfe of type a, then $\alpha\beta$ is an $\mathcal{L}_{T^s}$-wfe of type b.
(6) If α is an $\mathcal{L}_{T^s}$-wfe of type $\langle \langle s, t \rangle, t \rangle$ and β an $\mathcal{L}_{T^s}$-wfe of type t, then $\alpha\beta$ is an $\mathcal{L}_{T^s}$-wfe of type t.

It's obvious that $\mathcal{L}_T$ can be embedded in $\mathcal{L}_{T^s}$: Let ρ be the function from $\mathcal{L}_T$-wfe's to $\mathcal{L}_{T^s}$-wfe's that maps each constant, variable, and connective to itself, that maps $\mathcal{L}_T$-wfe's $\alpha\beta$ to $\rho(\alpha)\rho(\beta)$, and that maps $\mathcal{L}_T$-wfe's $\exists v \phi$, with $v \in V_a$, to $\exists^a v \rho(\phi)$.

Still taking models to be pairs $\mathfrak{M} = \langle D, \mathcal{I} \rangle$, we can define extensions of $\mathcal{L}_{T^s}$-wfe's relative to models and assignments in the by now familiar way. Clauses 1–4 are lexical entries, and clauses 5 and 6 composition rules.

---

[14] We must show that for every g and α, $[\![\alpha]\!]_{\lambda,s}^g = [\![\kappa(\alpha)]\!]_{\lambda,s}^g$. The interesting case is that of λ-abstraction. Suppose α is of type a and v of type b. Then $[\![(\lambda v \alpha)]\!]_\lambda^g = \underline{\lambda}r \in D_b. [\![\alpha]\!]_\lambda^{g[v:=r]}$, and $\kappa((\lambda v \alpha)) = (\lambda_b^a v \kappa(\alpha))$. Thus

$$[\![\kappa((\lambda v \alpha))]\!]_{\lambda,s}^g = [\![(\lambda_b^a v \kappa(\alpha))]\!]_{\lambda,s}^g = [\![\lambda_b^a v]\!]_{\lambda,s}^g ([\![\kappa(\alpha)]\!]_{\lambda,s}) = \underline{\lambda}r \in D_b. ([\![\kappa(\alpha)]\!]_{\lambda,s}(g[v:=r])) =$$

$$\underline{\lambda}r \in D_b. [\![\kappa(\alpha)]\!]_{\lambda,s}^{g[v:=r]}$$

from which the claim follows since the identity of $[\![\kappa(\alpha)]\!]_{\lambda,s}^{g[v:=r]}$ with $[\![\alpha]\!]_\lambda^{g[v:=r]}$ is guaranteed by the inductive hypothesis.

16 OPERATORS vs QUANTIFIERS

(1) $[\![c]\!]^g_s = \mathcal{I}(c)$ for $c \in \mathcal{C}_a$.
(2) $[\![v]\!]^g_s = g(v)$ for $v \in V_a$.
(3) $[\![\neg]\!]^g_s = \underline{\lambda}n \in \{0,1\}.\, 1-n$ and
    $[\![\wedge]\!]^g_s = \underline{\lambda}n \in \{0,1\}.\, \underline{\lambda}m \in \{0,1\}.\, \min\{n,m\}$.
(4) For $a \in \mathbf{T}$ and $v \in V_a$, $[\![\exists^a v]\!]^g_s = \underline{\lambda}f \in D_{\langle s,t\rangle}.\, \max\{f(g[v:=r]) \mid r \in D_a\})$.
(5) (EFA) If α is an $\mathcal{L}_{\mathbf{T}^s}$-wfe of type $\langle a, b\rangle$ and β an $\mathcal{L}_{\mathbf{T}^s}$-wfe of type a, then
    $[\![\alpha\beta]\!]^g_s = [\![\alpha]\!]^g_s([\![\beta]\!]^g_s)$.
(6) (IFA) If α is an $\mathcal{L}_{\mathbf{T}^s}$-wfe of type $\langle\langle s,t\rangle, t\rangle$ and β an $\mathcal{L}_{\mathbf{T}^s}$-wfe of type t, then
    $[\![\alpha\beta]\!]^g_s = [\![\alpha]\!]^g_s([\![\beta]\!]_s)$.

Readers will have no trouble establishing that ρ is an extensionally and intensionally faithful embedding of $\mathcal{L}_\mathbf{T}$ in $\mathcal{L}_{\mathbf{T}^s}$. Thus we now have a fully categorematic semantics for a λ-free language that embeds the original type-theoretical language $\mathcal{L}_\mathbf{T}$ in a faithful way.[15]

In this setting, $\exists^a v$ is *literally* an intensional operator, since its extension at any assignment takes assignment intensions as arguments. Indeed it's a strongly intensional operator, for the value of $[\![\exists^a v]\!]^g_s$ for some argument intensions does not depend only on those intensions' values at g.[16]

In $\mathcal{L}^\lambda_{\mathbf{T}^s}$, on the other hand, as in $\mathcal{L}^\lambda_\mathbf{T}$, $\exists^a$ is not an intensional operator, but rather an ordinary higher order function, as predicted by Frege and generalized quantifier theory. In particular, for any $a \in \mathbf{T}$, $\exists^a$ is of type $\langle\langle a,t\rangle, t\rangle$, which is again in $\mathbf{T}$. Nevertheless, even within $\mathcal{L}^\lambda_{\mathbf{T}^s}$ we have intensional operators, to wit, the $\lambda^t_a v$, which are of type $\langle\langle s,t\rangle,\langle a,t\rangle\rangle$. Indeed, the lambda abstractors λ^t_a are strongly intensional operators, for there are models $\mathfrak{M}$ and assignments g such that the value of $[\![\lambda^t_a v]\!]^g_{\lambda,s}$ for an intension as argument doesn't just depend on that intension's value at g.[17] But this means that, if we consider the compound operator $\exists^a \circ \lambda^t_a v$, which is arguably the counterpart of $\exists^a v$ in $\mathcal{L}_{\mathbf{T}^s}$, we find this 'quantifier' to be an intensional operator, indeed a strongly intensional operator, *stricto sensu*.

There thus seems to be considerable support for the view that quantifiers are intensional operators. But before we endorse that view, we should ask whether the formulations of type theory that most strongly underwrite it, $\mathcal{L}_{\mathbf{T}^s}$ and $\mathcal{L}^\lambda_{\mathbf{T}^s}$, are in fact theoretically satisfactory.

---

[15]Something like the approach just outlined is suggested by Levin (1988).
[16]Let D be the two-element set $\{x, y\}$, let the type a be e, let P and Q be of type $\langle e, t\rangle$, suppose that $\mathcal{I}(P)$ is the constant function on D with value 0, and that $\mathcal{I}(Q)$ maps x to 1 but y to 0. Let g be an assignment mapping every variable of type e to y. The value of $[\![\exists^a v]\!]^g_s$ for an intension as argument cannot in general depend only upon that intension's value at g, for $[\![\exists^a v]\!]^g_s([\![Pv]\!]_s) = 0 \neq 1 = [\![\exists^a v]\!]^g_s([\![Qv]\!]_s)$ even though $[\![Pv]\!]^g_s = 0 = [\![Qv]\!]^g_s$.
[17]Consider our previous example. Note that $[\![\lambda^e v]\!]^g_{\lambda,s}([\![Qv]\!]_{\lambda,s}) \neq [\![\lambda^e v]\!]^g_{\lambda,s}([\![Pv]\!]_{\lambda,s})$ but $[\![Pv]\!]^g_{\lambda,s} = [\![Qv]\!]^g_{\lambda,s}$.

4. What price categorematicity?

Let's take a step back and assess the price we've paid for eliminating syncategorematicity from the semantics of type theory – and what we've gotten in return.

$\mathcal{L}_{T^s}^{\lambda}$ and $\mathcal{L}_{T^s}$ each employ two formally distinct functional application rules in their semantics, namely extensional functional application, according to which the extension of $(\alpha\beta)$ at g is the result of applying the *extension* of α at g to the *extension* of β at g, and intensional functional application, according to which the extension of $(\alpha\beta)$ at g is the result of applying the *extension* of α at g to the *assignment intension* of β. This might strike one as problematic, especially given our earlier worries regarding the compositionality of theories with syncategorematic rules. For we again have *two* semantic rules – (EFA) and (IFA) – corresponding to *one* syntactic rule: concatenation of α and β. Moreover, we now have a language in which an expression stands at times for its extension and at times for its intension[18] – a feature that hasn't always sat well with philosophers of logic.[19]

Technically speaking, this feature of $\mathcal{L}_{T^s}^{\lambda}$ and $\mathcal{L}_{T^s}$ is a design artifact. As mentioned in footnote 1, we've here followed the principle adopted by von Fintel and Heim (2011), namely to minimize intensionality and retain extensional functional application wherever possible. We could, however, intensionalize the system in the manner suggested by von Fintel and Heim (2011, 12) and thereby do away with the *extensional* functional application rule. To illustrate, we could change the extension of, for example, negation at g to $[\![\neg]\!]_s^g = \lambda f \in D_{\langle s,t \rangle}. (1 - f(g))$, which would put it into type $\langle\langle s, t \rangle, t\rangle$.[20] Importantly, negation would now also be governed by the intensional functional application rule, since $[\![\neg\phi]\!]^g = [\![\neg]\!]^g([\![\phi]\!])$, and extensional functional application is no longer needed.

---

[18]The wfe (Pv), as it occurs in (¬(Pv)), contributes to the extension of (¬(Pv)) at g a truth value, but as a constituent of ($\exists^e$v(Pv)), (Pv) contributes to the extension at g of ($\exists^e$v(Pv)) an assignment intension.

[19]Thus Wittgenstein (1922): 'Two different symbols can therefore have the sign (the written sign or the sound sign) in common – they then signify in different ways' (TLP 3.321). 'Thus there easily arise the most fundamental confusions [...]' (TLP 3.324). 'In order to avoid these errors, we must employ a symbolism which excludes them, by not applying the same sign in different symbols [...] A symbolism, that is to say, which obeys the rules of logical grammar – of logical syntax [...]' (TLP 3.325). Frege, on the other hand, did allow for a sign to signify in different ways, e.g. to stand for its reference in ordinary contexts but for its sense in indirect contexts (Frege 1892). His treatment of identity in *Begriffsschrift* (1879, §8) in fact requires such a feature, since expressions are stipulated to refer to themselves when flanking the identity sign but to their ordinary referents when occurring elsewhere. See Pardey and Wehmeier (forthcoming) for details.

[20]Negation would then be a weakly intensional operator: Its extension at g would take intensions as arguments, but only the argument intension's value at g would matter to the function value assigned by the operator's extension.

So we can fiddle with the design to remove the awkwardness of having two functional application rules. There is a drawback to the fiddling, however, for it eliminates extensional operators altogether.[21] The distinction between extensional (*de iure* substitutive, as it were) and weakly intensional (*de facto* substitutive) operators that we can draw in $\mathcal{L}_{T^s}$ and $\mathcal{L}_{T^s}^\lambda$ disappears in the revised semantics, which seems like a conceptual disadvantage. But it's not clear how much weight such considerations ultimately carry.

There is a more serious problem, which might be called *syntactic pollution* of the semantic ontology. We set out to describe a simple non-linguistic structure, the extensional type hierarchy over a base set D. As long as we're willing to employ some syncategorematic means of variable-binding, as in $\mathcal{L}_T$ and $\mathcal{L}_T^\lambda$, we can accomplish this description by assigning as extensions to our wfe's objects naturally occurring in this type hierarchy: elements of D_e, D_t, and function spaces arising from these. But as soon as we insist on lexical entries for all variable-binders, it turns out that this ontology does not suffice: We must expand the type hierarchy over D by allowing 'conditionalization' on the set $\aleph_D$ of assignments. That is, in order to achieve a fully categorematic semantics for a language describing the extensional type hierarchy, we must assign to some expressions extensions that live outside that type hierarchy; indeed, extensions that essentially involve *syntactic* material, *viz*. the variables. In the case of $\mathcal{L}_{T^s}$, the quantifiers live in $D_{\langle\langle s,t\rangle,t\rangle}$ and hence outside $(D_a)_{a\in T}$; in the case of $\mathcal{L}_{T^s}^\lambda$, the abstractors λ_b^a live in $D_{\langle\langle s,a\rangle,\langle b,a\rangle\rangle}$, and hence likewise outside $(D_a)_{a\in T}$. Our semantic ontology has been contaminated with syntax.[22]

Rejection of syntactic contamination is not merely a philosophical matter of ontological purity; there are 'practical' consequences to constructing extensions out of assignments. To see this, let $\mathcal{L}'_{T^s}$ be exactly like $\mathcal{L}_{T^s}$ except that a particular variable w of $\mathcal{L}_{T^s}$ is replaced by a new variable w'. Fix a model $\mathfrak{M}$. Then $\mathcal{L}_{T^s}$ and $\mathcal{L}'_{T^s}$ have no compositional meanings (assignment intensions) in common, because these compositional meanings are functions on assignments, and $\mathcal{L}_{T^s}$ and $\mathcal{L}'_{T^s}$ have no assignments in common. Now suppose we want to analyze a fragment of English by

---

[21] Negation, for example, which is extensional in $\mathcal{L}_T$, $\mathcal{L}_T^\lambda$, $\mathcal{L}_{T^s}$ and $\mathcal{L}_{T^s}^\lambda$, because it takes truth values as arguments, becomes (weakly) intensional when we tweak its semantics in the way just indicated.

[22] Objections to this kind of contamination are, I take it, part of the motivation for the program of variable-free semantics – as Jacobson (2003, 58) points out, one wants the 'meaning of any linguistic expression [to be] simply some normal, healthy model-theoretic object – something constructed only out of the "stuff" that any theory presumably needs: individuals, worlds, times, perhaps events, etc'. Indeed, building extensions out of syntactic material seems to violate the basic idea of referential semantics, by making concrete representations (here, variables) essential to denotations.

way of indirect interpretation through a type-theoretical language. Then the compositional meanings assigned by this method to English phrases will differ depending on our choice between $\mathcal{L}_{T^s}$ and $\mathcal{L}'_{T^s}$. But surely this choice is completely arbitrary and shouldn't affect which model-theoretic object is assigned to a natural language expression as its meaning. Something has gone badly wrong: The meanings assigned are not, in fact, *purely* model-theoretic objects, but rather depend for their identity on the variables present in the underlying language.

It is sometimes suggested that syntactic pollution can be avoided by conceiving of a type a assignment $g_a : V_a \to D_a$ as the composition $\delta_a \circ \varepsilon_a$ of a bijection $\varepsilon_a : V_a \approx \mathbb{N}$ between the variables in V_a and the natural numbers $\mathbb{N}$ with a sequence $\delta_a : \mathbb{N} \to D_a$. The job originally carried out by the variables in V_a is then effectively taken over by their proxies modulo ε_a in $\mathbb{N}$, and the job of variable assignments is effectively taken over by the sequences $\delta_a : \mathbb{N} \to D_a$.[23] Accordingly we can, as it were, let the type s consist not of assignments $g = \bigcup \{g_a | a \in \mathbf{T}\}$ but rather of unions $\delta = \bigcup \{\delta_a | a \in \mathbf{T}\}$ over families $(\delta_a)_{a \in \mathbf{T}}$ of sequences $\delta_a : \mathbb{N} \to D_a$. Then the denotations that fall outside of the original extensional type hierarchy are built not out of syntactic material, but simply out of ordinary set-theoretical constructs, *viz.* natural numbers – or so the argument goes.

One way to see that this is no solution to the contamination problem is by observing that the proposal essentially requires taking the natural numbers to *be* the variables: the entities officially called variables that are bijectively correlated with the natural numbers are completely idle in the semantics, it's the correlated numbers that are doing all the work; hence the natural numbers are the *de facto* variables of such a system. But then the natural numbers *are* syntactic material, and the fact that meanings are constructed out of the natural numbers does nothing to mitigate the syntactic pollution issue.

Another way to reach the same conclusion is to notice that the proposal rules out by mere *fiat* using other countably infinite sets, such as the rational numbers $\mathbb{Q}$ or the integers $\mathbb{Z}$, as bijective correlates (proxies) of variables, and constructing meanings out of functions from *these* proxy sets into the relevant type hierarchy. If these other proxy sets are allowed, meanings depend on the choice of proxy set, just as in the orig-

---

[23]Heim and Kratzer (1998, 111 and 213) take this route, though to be sure they do not claim that they thereby avoid syntactic pollution.

inal set-up they depend on the choice of variables. But there is no principled reason to disallow $\mathbb{Q}$ or $\mathbb{Z}$, or any other countably infinite set, since there is nothing canonical about the choice of the natural numbers as proxies for variables. That is to say, one can obviously use a type-theoretical language based on rational or integer rather than natural number proxies for variables in the service of indirectly interpreting a fragment of English, and such a choice will materially affect which model-theoretic objects are assigned to English expressions as their meanings. This, however, clearly should not happen.

5. Fregean type theory

None of the versions of type theory we've canvassed appears to be entirely satisfactory: On the one hand, we have versions that bite the syncategorematic bullet and place all extensions within the extensional type hierarchy, but at the sacrifice of leaving variable-binders without lexical entries, thus arguably violating both Compositionality and Frege's Conjecture. On the other hand, we have versions that provide lexical entries for all operators, including the variable-binders, and satisfy both Compositionality and Frege's Conjecture, but at the sacrifice of contaminating the type-theoretic hierarchy with syntactic material, thereby making meanings language-dependent.

In this section, I argue that there is an attractive alternative version of extensional type theory that is superior to the systems considered so far in that it avoids both syncategorematicity *and* syntactic pollution. This alternative is extensional type theory in the spirit of Frege's *Grundgesetze* (1893).

Fregean type theory works with the original types in **T** and places its extensions within the extensional type hierarchy $(D_a)_{a \in \mathbf{T}}$ over any given model's domain D. The primitive symbols of the Fregean language $L_{\mathbf{T},F}$ are those of $\mathcal{L}_{\mathbf{T}^s}$ plus, for each $a \in \mathbf{T}$, a marker ξ^a for gaps of type a. The wfe's of $L_{\mathbf{T},F}$ and their types are defined as follows.

(1) Every $c \in \mathcal{C}_a$ is an $L_{\mathbf{T},F}$-wfe of type a.
(2) The connectives $\neg$ and $\wedge$ are $L_{\mathbf{T},F}$-wfe's of types $\langle t, t \rangle$ and $\langle t, \langle t, t \rangle \rangle$, respectively.
(3) For each $a \in \mathbf{T}$, the existential quantifier $\exists^a$ of type a is an $L_{\mathbf{T},F}$-wfe of type $\langle \langle a, t \rangle, t \rangle$.
(4) If α is an $L_{\mathbf{T},F}$-wfe of type $\langle a, b \rangle$ and β an $L_{\mathbf{T},F}$-wfe of type a, and no gap marker occurs in either α or β, then $\alpha\beta$ is an $L_{\mathbf{T},F}$-wfe of type b.

(5) If α is an $L_{T,F}$-wfe of type t, $c \in \mathcal{C}_a$ occurs in α, and no gap marker occurs in α, then $\alpha_c[\xi^a]$ is an $L_{T,F}$-wfe of type $\langle a, t \rangle$.[24]

(6) If π is an $L_{T,F}$-wfe of type $\langle a, t \rangle$ in which ξ^a occurs, and $v \in V_a$ doesn't occur in π, then $\exists^a v \pi_{\xi^a}[v]$ is an $L_{T,F}$-wfe of type t.

We next define the *Fregean extension* $[\![\alpha]\!]_F^{\mathfrak{M}}$ of an $L_{T,F}$-wfe α relative to a model $\mathfrak{M} = \langle D, \mathcal{I} \rangle$. Clauses 1–3 are lexical entries, clauses 4 and 5 are composition rules, and clause 6 is a *sui generis* rule corresponding to the syntactic operation of deletion.

(1) For $c \in \mathcal{C}_a$, $[\![c]\!]_F^{\mathfrak{M}} = \mathcal{I}(c)$.
(2) $[\![\neg]\!]_F^{\mathfrak{M}} = \lambda n \in \{0, 1\}.\ 1 - n$ and
$[\![\wedge]\!]_F^{\mathfrak{M}} = \lambda n \in \{0, 1\}.\ \lambda m \in \{0, 1\}.\ \min\{n, m\}$.
(3) For $a \in \mathbf{T}$, $[\![\exists^a]\!]_F^{\mathfrak{M}} = \lambda f \in D_{\langle a, t \rangle}.\ \max\{f(u) \mid u \in D_a\}$.
(4) $[\![\alpha\beta]\!]_F^{\mathfrak{M}} = [\![\alpha]\!]_F^{\mathfrak{M}}([\![\beta]\!]_F^{\mathfrak{M}})$.
(5) $[\![\exists^a v \pi_{\xi^a}[v]]\!]_F^{\mathfrak{M}} = [\![\exists^a]\!]_F^{\mathfrak{M}}([\![\pi]\!]_F^{\mathfrak{M}})$.
(6) $[\![\alpha_c[\xi^a]]\!]_F^{\mathfrak{M}} = \lambda u \in D_a.\ [\![\alpha]\!]_F^{\mathfrak{M}_c^u}$, where $\mathfrak{M}_c^u$ is the model that is just like $\mathfrak{M}$ except that it interprets the constant c as the object u.

Some comments are in order.

First, every *operator* of $L_{T,F}$ has its own lexical entry. This is so notwithstanding the fact that there is no lexical entry for the gap markers; these are obviously not operators in any syntactic sense.[25]

Second, the only semantic operation corresponding to syntactic modes of composition *stricto sensu* in $L_{T,F}$ is functional application; this is so notwithstanding the fact that (i) the syntactic operation of deletion does *not* correspond to functional application (deletion is, after all, not a mode of *composition*) and (ii) the left-hand sides of clauses 4 and 5 look somewhat different (their right-hand sides clearly show them both to be rules of functional application).[26] The semantics for $L_{T,F}$ is thus arguably in accord with Frege's Conjecture.

Third, all Fregean extensions in a model live in the extensional type hierarchy over the model's domain. Indeed, variable assignments play no role in the semantics at all. In particular, the quantifiers are obviously not

---

[24] In general, $\alpha_x[\beta]$ is the result of replacing each occurrence of x in α with β. Thus $\alpha_c[\xi^a]$ is the result of replacing each occurrence of c in α with the gap marker ξ^a of type a. More poetically, it is the result of erasing every occurrence of c in α, leaving behind an expression with gaps fit to receive wfe's of type a.

[25] In fact, if I may be allowed to wax metaphorical, they are mere *absences* of syntactic material.

[26] The difference in appearance of the left-hand sides is due merely to the fact that quantifications require additional punctuation (provided by the variable) so that the predicate argument be uniquely reconstructible from the quantified sentence.

intensional operators, since their extensions are ordinary higher order functions located within the extensional types **T**.

Fourth, if we define the *Fregean intension* $[\![\alpha]\!]_F$ of an $L_{T,F}$-wfe α as the function on the class of all models that maps each model $\mathfrak{M}$ to the Fregean extension $[\![\alpha]\!]_F^{\mathfrak{M}}$ of α in $\mathfrak{M}$, Fregean intensions are compositional.[27]

If desired, lambda abstraction can easily be integrated into Fregean type theory. Let the primitive symbols of the lambda-enriched Fregean language $L_{T,F}^{\lambda}$ be those of $L_{T,F}$ together with the symbol λ_a^b for each pair $\langle a, b \rangle \in \mathbf{T} \times \mathbf{T}$, as well as parentheses. The $L_{T,F}^{\lambda}$-wfe's are defined as follows:

(1) Every $c \in \mathcal{C}_a$ is an $L_{T,F}^{\lambda}$-wfe of type a.
(2) The connectives $\neg$ and $\wedge$ are $L_{T,F}^{\lambda}$-wfe's of types $\langle t, t \rangle$ and $\langle t, \langle t, t \rangle \rangle$, respectively.
(3) For each $a \in \mathbf{T}$, the existential quantifier $\exists^a$ is an $L_{T,F}^{\lambda}$-wfe of type $\langle \langle a, t \rangle, t \rangle$.
(4) For each $\langle a, b \rangle \in \mathbf{T} \times \mathbf{T}$, the abstractor λ_a^b is an $L_{T,F}^{\lambda}$-wfe of type $\langle \langle a, b \rangle, \langle a, b \rangle \rangle$.
(5) If α is an $L_{T,F}^{\lambda}$-wfe of type $\langle a, b \rangle$ and β an $L_{T,F}^{\lambda}$-wfe of type a, and no gap marker occurs in either α or β, then $(\alpha\beta)$ is an $L_{T,F}^{\lambda}$-wfe of type b.
(6) If α is an $L_{T,F}^{\lambda}$-wfe of type b and $c \in \mathcal{C}_a$ occurs in α, and no gap marker occurs in α, then $\alpha_c[\xi^a]$ is an $L_{T,F}^{\lambda}$-wfe of type $\langle a, b \rangle$.
(7) If π is an $L_{T,F}^{\lambda}$-wfe of type $\langle a, b \rangle$ in which ξ^a occurs, and $v \in V_a$ doesn't occur in π, then $\lambda_a^b v \pi_{\xi^a}[v]$ is an $L_{T,F}^{\lambda}$-wfe of type $\langle a, b \rangle$.

Needless to say, existential quantification can be handled via clause 5, $L_{T,F}$-wfe's $\exists^a v \phi_{\xi^a}[v]$ being replaced by $L_{T,F}^{\lambda}$-wfe's $(\exists^a(\lambda_a^t v \phi_{\xi^a}[v]))$.

Fregean extensions $[\![\alpha]\!]_{F,\lambda}^{\mathfrak{M}}$ for $L_{T,F}^{\lambda}$-wfe's α relative to a model $\mathfrak{M}$ are defined as follows. Clauses 1–4 are lexical entries, clauses 5 and 6 are composition rules and clause 7 is a *sui generis* rule corresponding to the syntactic operation of deletion.

(1) For $c \in \mathcal{C}_a$, $[\![c]\!]_{F,\lambda}^{\mathfrak{M}} = \mathcal{I}(c)$.
(2) $[\![\neg]\!]_{F,\lambda}^{\mathfrak{M}} = \underline{\lambda} n \in \{0,1\}. 1-n$ and
    $[\![\wedge]\!]_{F,\lambda}^{\mathfrak{M}} = \underline{\lambda} n \in \{0,1\}. \underline{\lambda} m \in \{0,1\}. \min\{n,m\}$.
(3) For $a \in \mathbf{T}$, $[\![\exists^a]\!]_{F,\lambda}^{\mathfrak{M}} = \underline{\lambda} f \in D_{\langle a, t \rangle}. \max\{f(u) \mid u \in D_a\}$.
(4) For $a, b \in \mathbf{T}$, $[\![\lambda_a^b]\!]_{F,\lambda}^{\mathfrak{M}} = \mathrm{Id}_{\langle a, b \rangle}$ (the identity function on $D_{\langle a, b \rangle}$).
(5) $[\![(\alpha\beta)]\!]_{F,\lambda}^{\mathfrak{M}} = [\![\alpha]\!]_{F,\lambda}^{\mathfrak{M}}([\![\beta]\!]_{F,\lambda}^{\mathfrak{M}})$.
(6) $[\![\lambda_a^b v \pi_{\xi^a}[v]]\!]_{F,\lambda}^{\mathfrak{M}} = [\![\lambda_a^b]\!]_{F,\lambda}^{\mathfrak{M}}([\![\pi]\!]_{F,\lambda}^{\mathfrak{M}})$.

---

[27] This follows *mutatis mutandis* from results of Wehmeier (2018).

(7) $[\![\alpha_c[\xi^a]]\!]_{F,\lambda}^{\mathfrak{M}} = \underline{\lambda} u \in D_a.\ [\![\alpha]\!]_{F,\lambda}^{\mathfrak{M}_c^u}.$

Quantification now falls under the functional application clause 5; but lambda abstraction, too, is handled by functional application, as is evident from the right-hand side of clause 6, notwithstanding the fact that the lambda-abstractor is interpreted trivially as the identity function of the relevant type.[28]

Now define the Fregean intension $[\![\alpha]\!]_{F,\lambda}$ of an $L_{T,F}^{\lambda}$-wfe α as the function on the class of all models that maps each model $\mathfrak{M}$ to the Fregean extension $[\![\alpha]\!]_{F,\lambda}^{\mathfrak{M}}$ of α in $\mathfrak{M}$. These Fregean intensions are compositional.[29]

In Fregean type theory, whether in the shape of $L_{T,F}$ or $L_{T,F}^{\lambda}$, we thus have

- a compositional semantics
- that abides by Frege's Conjecture
- in which all operators have lexical entries and
- whose ontology is not contaminated by syntactic material.

We do have a *sui generis* syntactic rule, deletion, that is distinct from concatenation, and corresponding to it a *sui generis* semantic rule that isn't functional application. But given the drawbacks of the alternatives – syncategorematic semantic clauses and concomitant violations of compositionality on the one hand, syntactic contamination of the semantic ontology on the other – the Fregean theories seem preferable to any of the more familiar type theories we've examined.

Moreover, the Fregean has an intuitive story to tell about the new syntactic and semantic rules: How do we arrive at the complex predicate *respects Laszlo and despises Strasser*? We first construct the sentence *Rick respects Laszlo and Rick despises Strasser*, then delete the name *Rick* to obtain *ξ respects Laszlo and ξ despises Strasser*. Finally we apply λ and get the one-place complex predicate λx. *x respects Laszlo and x despises Strasser*. How do we compute that predicate's extension? It's identical to that of *ξ respects Laszlo and ξ despises Strasser*, which we find by taking any sentence from which it could've been obtained by deletion, say *Rick respects Laszlo and Rick despises Strasser*, and going through all the logically

---

[28]Frege (1893) employs a device similar to lambda abstraction: the value-range operator. This operator is *not* interpreted trivially, as it is intended to inject $D_{\langle e,e \rangle}$ into D_e, thereby engendering inconsistency (see Wehmeier 2004). The lambda-operator, despite being *interpreted* trivially, is not itself trivial, as it enables complex predicates to figure as functors or arguments in the syntactic concatenation rule 5.
[29]Compare again Wehmeier (2018).

possible extensions of the deleted name. Any object that, taken as the extension of the name, makes the sentence true, goes into the predicate's extension; any object that, taken as the name's extension, makes the sentence false, stays out of the predicate's extension. Thus Richard Blaine is in, for when the name *Rick* refers to Blaine (as it in fact does), *Rick respects Laszlo and Rick despises Strasser* is true; on the other hand, Major Strasser's aide Heinz is out, since *Rick respects Laszlo and Rick despises Strasser* is false when *Rick* refers to Heinz.

Crucially for our purposes, in neither Fregean theory do we have *any* intensional operators. This is trivially true on one reading of the claim: a Fregean model only contains entities living in one of the extensional types, where functions defined on variable assignments *qua* indices simply don't occur; accordingly, no operator extension can be a function defined on such indices.

The claim is still trivially true if we relax the definition of intensional operator to allow syncategorematic semantic clauses structurally identical to those for the modal operators; this is because the Fregean theories don't employ any syncategorematic operator clauses.

But the claim is *still* true on a broader interpretation of 'intension' as encompassing functions whose domain is the class of all models (taken as surrogates for indices): The Fregean extension, in any given model, of any operator in $L_{T,F}$ or $L_{T,F}^\lambda$, takes as arguments entities living in the extensional type hierarchy over the model's domain; in other words, no operator extension ever takes as arguments functions whose domain is the class of all models. There is thus *no* sense in which $L_{T,F}$ or $L_{T,F}^\lambda$ contain intensional operators.[30]

6. Closing

If Fregean versions of type theory are indeed superior to the other formulations we have surveyed, we should presumably look to $L_{T,F}$ or $L_{T,F}^\lambda$ when trying to understand the semantic status of quantifiers. And as we observed at the end of the previous section, there is no sense in which the quantifiers appear as intensional operators in $L_{T,F}$ or $L_{T,F}^\lambda$; not, in any case, if we define intensional operators the way we did at the beginning of this paper.

---

[30]Moreover, neither the quantifiers nor the lambda abstractors satisfy even the syntactic criterion for intensional operators, to wit, that they operate on sentences: The syntactic arguments of $\exists^a$ in $L_{T,F}$ are *never* sentential, but rather explicitly predicative, in that they must contain a gap marker and have references that live in predicative types of the form $\langle a, t \rangle$, and likewise for the syntactic arguments of λ_a^r in $L_{T,F}^\lambda$.

That definition derives from the familiar possible-world treatment of modal and temporal operators. While it has long been a commonplace that these operators *can be construed as* quantifiers over possible worlds or times, linguistic semantics seems to take more and more seriously the notion that they literally *are* quantifiers.[31] But if modals and tenses are actually quantifiers, and quantifiers aren't intensional operators, as I've argued here, then intensional operators are uninteresting, merely theoretical constructs to which no feature of natural language answers.

Acknowledgements

Thanks to the editors of this special issue and an anonymous referee, as well as Greg Scontras, for helpful comments and suggestions. This paper was presented at the Center for Language, Information and Philosophy (CLIP) of the University of Cologne, Germany, and at the Department of Linguistics of the University of Frankfurt, Germany, both in October of 2017. I am grateful to the audiences for valuable feedback, especially Daniel Gutzmann, Stefan Hinterwimmer, and Ede Zimmermann. Funding from the National Endowment for the Humanities is gratefully acknowledged.

Disclosure statement

No potential conflict of interest was reported by the author.

Funding

Funding was provided by the National Endowment for the Humanities (Grant No. FA-232235-16). Any views, findings, conclusions, or recommendations expressed in this article do not necessarily represent those of the National Endowment for the Humanities.

ORCID

Kai F. Wehmeier http://orcid.org/0000-0002-6468-497X

References

Belnap, N. 2005. "Under Carnap's Lamp: Flat Pre-semantics." *Studia Logica* 80: 1–28.
von Fintel, K., and I. Heim. 2011. *Intensional Semantics*. Spring 2011 ed. Cambridge, MA: MIT. Unpublished lecture notes.
Frege, G. 1879. *Begriffsschrift*. Halle: Louis Nebert. Translated as *Concept Script* by S. Bauer–Mengelberg in J. van Heijenoort (ed.), *From Frege to Gödel: A Source Book in Mathematical Logic 1879–1931*. Cambridge, MA: Harvard University Press, 1967.

---

[31] See e.g. King (2003), von Fintel and Heim (2011, §8.2) and Schaffer (2012).

Frege, G. 1892. "Über Sinn und Bedeutung." *Zeitschrift für Philosophie und philosophische Kritik* NF 100: 25–50. Translated as "On Sense and Reference" by M. Black in *Translations from the Philosophical Writings of Gottlob Frege*, edited and translated by P. Geach and M. Black, 3rd ed. Oxford: Blackwell, 1980.

Frege, G. 1893. *Grundgesetze der Arithmetik. Vol. I.* Jena: Hermann Pohle. English translation in *Gottlob Frege: Basic Laws of Arithmetic*, edited and translated by P. Ebert and M. Rossberg, Oxford: Oxford University Press, 2013.

Gamut, L. T. F. 1991. *Logic, Language, and Meaning, Vol. 2.* Chicago: University of Chicago Press.

Heim, I., and A. Kratzer. 1998. *Semantics in Generative Grammar.* Oxford: Blackwell.

Jacobson, P. 2003. "Binding Without Pronouns (and Pronouns Without Binding)." In *Resource-Sensitivity, Binding, and Anaphora*, edited by R. Oehrle and G.-J. Kruiff, 57–96. Dordrecht: Kluwer.

King, J. 2003. "Tense, Modality, and Semantic Values." *Philosophical Perspectives* 17 (Language and Philosophical Linguistics): 195–245.

Levin, H. 1988. "A Philosophical Introduction to Categorial and Extended Categorial Grammar." In *Categorial Grammar*, edited by W. Buszkowski, W. Marciszewski, and J. van Benthem, 173–195. Amsterdam: John Benjamins.

Pagin, P., and D. Westerstahl. 2010. "Compositionality I: Definitions and Variants." *Philosophy Compass* 5 (3): 250–264.

Pardey, U., and K. Wehmeier. Forthcoming. "Frege's *Begriffsschrift* Theory of Identity Vindicated." *Oxford Studies in Philosophy of Language.*

Rabern, B. 2012. "Monsters in Kaplan's Logic of Demonstratives." *Philosophical Studies* 164: 393–404.

Schaffer, J. 2012. "Necessitarian Propositions." *Synthese* 189 (1): 119–162.

Varzi, A. 1993. "Do We Need Functional Abstraction?" In *Philosophy of Mathematics – Proceedings of the 15th International Wittgenstein Symposium*, Part 1, edited by J. Czermak, 407–415. Vienna: Hölder-Pichler-Tempsky.

Wehmeier, K. 2004. "Russell's Paradox in Consistent Fragments of Frege's *Grundgesetze.*" In *One Hundred Years of Russell's Paradox: Proceedings of the 2001 Munich Conference*, edited by G. Link, 247–257. Berlin: de Gruyter.

Wehmeier, K. F. 2018. "The Proper Treatment of Variables in Predicate Logic." *Linguistics and Philosophy* 41 (2): 209–249.

Wittgenstein, L. 1922. *Tractatus Logico-Philosophicus.* London: Kegan Paul.

Binding bound variables in epistemic contexts

Brian Rabern

ABSTRACT
Quine insisted that the satisfaction of an open modalised formula by an object depends on how that object is described. Kripke's 'objectual' interpretation of quantified modal logic, whereby variables are rigid, is commonly thought to avoid these Quinean worries. Yet there remain residual Quinean worries for epistemic modality. Theorists have recently been toying with assignment-shifting treatments of epistemic contexts. On such views an epistemic operator ends up binding all the variables in its scope. One might worry that this yields the undesirable result that any attempt to 'quantify in' to an epistemic environment is blocked. If quantifying into the relevant constructions is vacuous, then such views would seem hopelessly misguided and empirically inadequate. But a famous alternative to Kripke's semantics, namely Lewis' counterpart semantics, also faces this worry since it also treats the boxes and diamonds as assignment-shifting devices. As I'll demonstrate, the mere fact that a variable is bound is no obstacle to binding it. This provides a helpful lesson for those modelling *de re* epistemic contexts with assignment sensitivity, and perhaps leads the way toward the proper treatment of binding in both metaphysical and epistemic contexts: Kripke for metaphysical modality, Lewis for epistemic modality.

It is now commonly acknowledged that much early theorising concerning modal notions suffered from various confusions and conflations. A major advance, at least in twentieth century philosophy, was Kripke's work, which brought great clarity to the nature of—and varieties of—modality (e.g. Kripke 1963, 1980). The background to much of Kripke's work in this area concerns issues in the model-theoretic semantics for modal logic, especially quantified modal logic and the forceful Quinean objections to such an enterprise (e.g. Quine 1947, 1953b). Quine insisted that quantifying into modal contexts was incoherent, since the truth of a formula ∃xFx would require the satisfaction of the embedded open formula □Fx by some object. But Quine insisted that it didn't make sense to say of an object that *it* must (or might) have some property, independently of how the object was described. For example, what could it mean for an object *a* to satisfy ⌜□(x > 7)⌝? It seems to matter whether *a* is described as 'the number 8' or as 'the number of planets'. Consider the contrast between (1) and (2):

(1) □**(the number of planets** > 7)
(2) □**(8** > 7)

When giving this objection, Quine understood 'necessity' in terms of *semantical* necessity, or analyticity, or apriority (as did the modal logicians he was most immediately reacting to, e.g. Ruth Barcan (1946) and Rudolf Carnap (1947)). In those terms, his point is that while it makes sense to say that 'the inventor of bifocals is the inventor of bifocals' is semantically necessary (i.e. analytic), it makes little sense to say that the inventor of bifocals (i.e. Ben Franklin) is such that it is analytic that *he* is the inventor of bifocals. If it makes any sense at all to say of an object that ⌜x is the inventor of bifocals⌝ is analytically true of it, it would seem to depend on how that object was described (e.g. 'the inventor of bifocals' versus 'the founder of the American Philosophical Society').

After distinguishing various notions of modality—epistemic, semantical, and metaphysical—Kripke responds to the Quinean worry by shifting the focus to the objective or metaphysical understanding of modality, whereby he insists that it makes perfect sense talk about what properties certain objects *had to have* had or what properties they *could have* lacked (see Burgess 1998). And he insists that the person who objects to these common sense notions along the Quinean lines is being a *philosopher* in the pejorative sense:

> Suppose that someone said, pointing to Nixon, 'That's the guy who might have lost'. Someone else says, 'Oh no, if you describe him as *Nixon*, then he might have lost; but, of course, describing him as *the winner*, then it is not true that he might have lost'. Now which one is being the philosopher, here, the unintuitive man? (Kripke 1980, 41)

In the background, Kripke has a model theory for quantified modal logic whereby a formula such as '□Fx' is true of an object independently of how it is described or denoted. This is due to the fact that on this, now standard, 'objectual' interpretation of quantified modal logic the value of a variable relative to an assignment is independent of the world parameter (variables are *rigid* de jure) and modals don't meddle with the assignment parameter.

To see this consider a standard version of modal predicate logic with identity. In addition to brackets, the basic symbols consist of the following:

Variables: $x, y, z, \ldots$
Predicates: $F, G, H, \ldots$
Constants: $\neg, \wedge, \exists, =, \Box$.

For any sequence of variables, $\alpha_1, \ldots, \alpha_n$, any *n*-place predicate π, and any variables a and β, the sentences of the language are provided by the following definition:

$$\phi ::= \pi\alpha_1 \ldots \alpha_n \mid \alpha = \beta \mid \neg\phi \mid (\phi \wedge \phi) \mid \exists \alpha \phi \mid \Box \phi.$$

Let a model $\mathfrak{A} = \langle W, R, D, I \rangle$, where W is a (non-empty) set of worlds, R is a binary relation on W (an accessibility relation), D is a (non-empty) set of individuals, and I is the interpretation, which relative to a world assigns the predicate sets of tuples of individuals drawn from D. We can then provide the semantic clauses relative to a variable assignment, a world, and a model $\mathfrak{A}$ as follows:

- $[\![\alpha]\!]^{g,w} = g(\alpha)$
- $[\![\pi\alpha_1 \ldots \alpha_n]\!]^{g,w} = 1$ iff $\langle [\![\alpha_1]\!]^{g,w}, \ldots, [\![\alpha_n]\!]^{g,w} \rangle \in I(\pi,w)$
- $[\![\alpha = \beta]\!]^{g,w} = 1$ iff $[\![\alpha]\!]^{g,w} = [\![\beta]\!]^{g,w}$
- $[\![\neg\phi]\!]^{g,w} = 1$ iff $[\![\phi]\!]^{g,w} = 0$
- $[\![\phi \wedge \psi]\!]^{g,w} = 1$ iff $[\![\phi]\!]^{g,w} = 1$ and $[\![\psi]\!]^{g,w} = 1$
- $[\![\exists\alpha\phi]\!]^{g,w} = 1$ iff for some g' that differs from g at most in that $g'(\alpha) \neq g(\alpha)$, $[\![\phi]\!]^{g',w} = 1$
- $[\![\Box\phi]\!]^{g,w} = 1$ iff for all w' such that wRw', $[\![\phi]\!]^{g,w'} = 1$.

Many have agreed with Kripke that this interpretation of quantified modal logic fits well with an 'objective' understanding of the modals, e.g. the kinds of readings of modals that linguistics might label nomological, historical, or circumstantial. Kripke's work ushered in intense philosophical and logical investigation of these objective modalities—with a special focus on the most general variety, which Kripke called *metaphysical* modality.

But, as is well known, there are other 'flavours' of modality (see Kratzer 1977). The standard semantics for quantified modal logic may not be the most appropriate for certain varieties of modality—in particular, the *epistemic* modalities. Philosophical discussions have tended to echo Kripke's stance on epistemic modality—it is often mentioned in order to distinguish it from genuine *metaphysical* modality, but is then quickly set aside as 'merely epistemic'.[1] In an unpublished note from the late 1990s, David Chalmers (unpublished) suggests that the post-Kripkean era has been complicit in a 'tyranny of the subjunctive'. He insists that the discussion has been overly biased toward metaphysical modality, and that epistemic necessity deserves the same focus and study as metaphysical necessity (see also Chalmers 2011). Chalmers invites us to imagine an alternative universe in which Kripke instead focused primarily on epistemic modality. Chalmers suggests that Kripke would have written a completely different book, perhaps called *Naming and possibility*, within which he, among other things would:

- defend a link between necessity and apriority,
- argue for a more descriptivist view of names,
- argue against the necessity of identity,
- have a different view about *de re* modality and quantified modal logic.

Even the actual Kripke admits that although identities are metaphysically necessary, they are not *epistemically* necessary. He maintains that 'for all we knew in advance, Hesperus wasn't Phosphorus' (Kripke 1980, 104), and that 'we do not know *a priori* that Hesperus is Phosphorus' (Kripke 1980, 104).[2] When discussing the sense in which prior to empirical investigation it could turn out either way whether Hesperus is Phosphorus, he says

> ... there is one sense in which things might turn out either way ... Obviously, the 'might' here is purely 'epistemic'—it merely expresses our present state of ignorance, or uncertainty. (Kripke 1980, 103)

---

[1]This attitude was probably more prevalent 10 years ago than it is today. There has been a recent flurry of interest in epistemic modalities mostly stemming from issues in the contextualism/relativism literature, see Egan and Weatherson (2011). Even here quantified epistemic modality or epistemic modality *de re* has not been of primary focus—yet important explorations in this vein include Gerbrandy (1997), Gerbrandy (2000), Aloni (2001), Aloni (2005), Yalcin (2015), Swanson (2010), Chalmers (2011), and Ninan (forthcoming), 'Quantification and Epistemic Modality'.

[2]Kripke in fact makes many statements about what *might* turn out (or might have turned out) in the epistemic sense, which are not possible in the metaphysical sense, e.g. he admits that Nixon might be an automaton (46), a certain wooden table might have been, given certain evidence, made of ice (145–146), and Gold might not to be an element (143), etc.: 'If I say, 'Gold *might* turn out not to be an element', I speak correctly; 'might' here is *epistemic* and expresses the fact that the evidence does not justify a priori (Cartesian) certainly that gold is an element' (Kripke 1980, 143fn72). See also Kripke (1971), footnote 15.

Although Kripke is downplaying the 'merely' epistemic possibility, he nevertheless seems to be committed to the claim that an agent could use an epistemic modal to truly say 'Hesperus might not be Phosphorus'. Agents can be ignorant of metaphysical necessities and can express such ignorance using epistemic modals. To get a solid example on the table consider this one:

> An ancient astronomer has been investigating the moons of various celestial bodies. She has gathered various evidence from observation and geometrical calculation. In the mornings, she has been investigating a celestial body that she calls 'Phosphorus'. The visibility in the morning is poor, but given certain evidence she suspects it has a moon. Also there is a celestial body that she calls 'Hesperus', which she has been investigating in the evenings. She has been able to collect an immense amount of evidence, and according to her best geometrical models it'd be inconsistent with the data for it to have a moon.

Given this situation, it would be felicitous for the astronomer to say each of the following:

(3) Phosphorus might have a moon
(4) Hesperus must not have a moon

And this is so even though Hesperus is identical to Phosphorus. But on the standard interpretation (3) and (4) cannot both be true, since the following sentences would have to be jointly satisfiable (where 'x' and 'y' are the two names of the planet, and 'M' is the predicate 'having a moon'):

$$x = y \quad \Diamond Mx \quad \neg \Diamond My.$$

But, of course, they can't be. To satisfy the latter two, it must be that

$$[\![\Diamond Mx]\!]^{g,w} \neq [\![\Diamond My]\!]^{g,w}.$$

That can only be if $g(x) \neq g(y)$, but to satisfy the first it must be that $g(x) = g(y)$.[3]

Returning to the discussion of Quine's objection to quantified modal logic, it seems that Kripke is right that the following is confused:

---

[3]Since we are presumably in a different epistemic state than this astronomer, some might insist that we should not accept her utterances. This feature is well known in the contextualism/relativism literature. To avoid this complication, it may be better to consider an example that avoids it. Consider utterances of 'Bansky might be Robert Del Naja' and 'Banksy might be Robin Gunningham'. It seems that both are true, but they can't be given and the standard semantic assumptions combined with the fact that Robert Del Naja is distinct from Robin Gunningham.

> [Winning the 1968 election] is a contingent property of Nixon only relative to our referring to him as 'Nixon'. But if we designate Nixon as 'the man who won the election in 1968', then it will be a [metaphysically] necessary truth, of course, that the man who won the election in 1968, won the election in 1968. (Kripke 1980, 40)

In other words, the relevant reading of

(5) The winner might not have been the winner

is perfectly fine. But altering it to epistemic modality gets the opposite result.

> Winning the 1968 election is not epistemically necessary of Nixon relative to our referring to him as 'Nixon'. But if we designate Nixon as 'the man who won the election in 1968', then it will be epistemically necessary, of course, that the man who won the election in 1968, won the election in 1968.

Or in other words, as Yalcin (2015) has emphasised, there is no true reading of the following epistemic claim:

(6) The winner might not be the winner

Whether or not a property holds of an object by epistemic possibility (or necessity) seems to depend on how that object is described. But this, of course, is in direct conflict with the standard interpretation of quantified modal logic. These issues concerning *epistemic* modality *de re*, in particular, are not really settled by Kripke's discussion—they are ignored, and thus in the epistemic case there remain residual Quinean worries (cf. Burgess 1998 and Chalmers 2011, 89). These questions remain: What is the status of epistemic modality *de re* and 'quantifying in' to epistemic contexts? What changes to the model theory are required to accommodate quantified epistemic modality? And most directly how should we interpret variables under epistemic modals so that '$x = y$' holds while '$\Box(x = y)$' doesn't?

In one form or another, these questions have been addressed since the mid-1900s, and I can't hope to answer them in a completely comprehensive and satisfactory way here. My modest aim here is to run a continuous thread through various existing strands that are already in the literature, highlight lessons along the way, and sketch out an appealing approach to quantified epistemic modality. I make no claim to novelty in detail—my contributions here involve summary, emphasis, and gestures toward new horizons.

Central to my discussion is the idea that certain modals ought to be understood as 'assignment-shifting' devices. Various theorists have been toying with assignment-shifting treatments of *epistemic contexts* such as attitude verbs and epistemic modals (e.g. Cumming 2008; Santorio 2012; Ninan 2012; Pickel 2015; Rieppel 2017). On such views, an epistemic '□' ends up binding the x in □Fx. One might worry that this kind of binding yields the undesirable result that any attempt to 'quantify in' to an epistemic environment is blocked, e.g. $\exists x$□Fx would be a case of vacuous quantification. If quantifying into the relevant constructions is vacuous, then such views would seem hopelessly misguided and empirically inadequate.[4]

Here there are enlightening, and perhaps surprising, connections to a famous alternative to Kripke's semantics for quantified modal logic, namely Lewis' counterpart semantics (Lewis 1968). In an important sense, these assignment-shifting treatments of modals just are versions of a counterpart semantics—or the other way round: Lewis' counterpart semantics treats the *boxes* and *diamonds* as assignment-shifting (in addition to worlds-shifting) devices. Thus similar worries about quantifying in and vacuity arise for Lewis' counterpart semantics—in the well-known footnote 13, just before giving the 'Humphrey objection', Kripke (1980) complains precisely about this by insisting that Lewis' view 'suffers from a purely formal difficulty'. But as I'll demonstrate below the mere fact that a variable is bound is no obstacle to binding it. This provides a helpful lesson for those modelling *de re* epistemic contexts with assignment sensitivity, and perhaps leads the way toward the proper treatment of binding in both metaphysical and epistemic contexts: *Kripke for objective modality, Lewis for epistemic modality*.

1. Binding the bound

Let's start with some folk wisdom concerning quantification and binding that turns out to be false—obviously false, perhaps, but the counterexample brings out the devices that may prove useful for modelling epistemic contexts. Consider this formula of first-order logic:

---

[4]One might insist, on the contrary, that blocking quantification into epistemic contexts is a feature, not a bug. One might insist that *de re* epistemic modality doesn't even make sense, or one might insist that there are certain constraints on the scopal interaction of quantifiers with epistemic modals which disallow quantifying in, see the Epistemic Containment Principle of Von Fintel and Iatridou (2003). While there may be certain kinds of constructions that don't allow for the quantifier to scope over the epistemic modal ('everyone inside might be outside'), it also seems clear that there are cases of quantifying in, e.g. after painting the ceiling I might warn you to walk carefully by saying: 'Almost every square inch of the floor might have paint on it' (Swanson 2010; see also Yalcin 2015). Or discussing the lottery I might say 'Every ticket is very likely to have lost' (cf. the cases pointed to in Lennertz 2015).

(7) ∀x∃xFx

This is a paradigm example of what is called 'vacuous quantification'. The occurrence of the variable 'x' in 'Fx' is already bound by '∃x' in the subformula '∃xFx', thus prefixing '∀x' is idle—the universal quantifier is *vacuous*. For example, the following is a theorem concerning vacuous quantification from introductory logic texts (see, e.g. Kalish and Montague 1964, 164–165).

(8) ∀x∃xFx ↔ ∃xFx

In general, one might insist on the following principle concerning binding and vacuity.

> THE PRINCIPLE OF VACUOUS QUANTIFICATION. If all the variables in a formula ϕ are bound, then for any quantifier Σ, $\Sigma\phi \leftrightarrow \phi$.

In slogan: *You can't bind a bound variable!* This is the bit of folk wisdom that isn't exactly correct. There is more going on with 'vacuous quantification' than is commonly recognised.

In order to show that the Principle of Vacuous Quantification is false I will introduce a simple language that sticks close to the syntax and semantics of first-order logic—this helps to demonstrate that there is nothing tricky going on in my counterexample.

Bear with me while I set out the formalities. First we define the syntax. In addition to parenthesis, the basic symbols consist of the following:

Variables: x, y, z, ...
Predicates: F, G, H, ...
Connectives: ¬, ∧
Quantifiers: ∃, ᴚ.

For any sequence of variables $\alpha_1, \ldots, \alpha_n$, any *n*-place predicate π, and any variable a, the sentences of the language are provided by the following grammar:

$$\phi ::= \pi\alpha_1 \ldots \alpha_n \mid \neg\phi \mid (\phi \wedge \phi) \mid \exists\alpha\phi \mid ᴚ\alpha\phi.$$

This language looks essentially like predicate logic, and we can define the other usual operators (e.g. '∨', '→', '∀', etc.) as abbreviations in terms of our basic symbols. The only novel thing about the language, thus far, is the addition of a new quantifier symbol 'ᴚ'. There is nothing interesting about it syntactically, and although it will be given an interpretation that is different from '∃' it is essentially a kind of existential quantifier.

Turning to the semantics, let a model $\mathfrak{A} = \langle D, I \rangle$, where D is a (non-empty) set of individuals and I is an interpretation function, which assigns values to the predicates. Since our language has variable-binding operators we relativise to an *assignment*, which assigns values to the variables. An assignment g is a function from the set of variables to set of individuals D. We provide the following recursive semantics, in the style of Tarski, by recursively defining 1 (truth or satisfaction) relative to an assignment g[5]:

- $[\![\alpha]\!]^g = g(\alpha)$
- $[\![\pi\alpha_1 \ldots \alpha_n]\!]^g = 1$ iff $\langle [\![\alpha_1]\!]^g, \ldots, [\![\alpha_n]\!]^g \rangle \in I(\pi)$
- $[\![\neg\phi]\!]^g = 1$ iff $[\![\phi]\!]^g = 0$
- $[\![\phi \wedge \psi]\!]^g = 1$ iff $[\![\phi]\!]^g = 1$ and $[\![\psi]\!]^g = 1$
- $[\![\exists\alpha\phi]\!]^g = 1$ iff for some g' that differs from g at most in that $g'(\alpha) \neq g(\alpha)$, $[\![\phi]\!]^{g'} = 1$
- $[\![⅄\alpha\phi]\!]^g = 1$ iff for some g' that differs from g at most in that $g'(\alpha) > g(\alpha)$, $[\![\phi]\!]^{g'} = 1$.

The last clause deserves comment, since it appeals to the *greater than* relation. For this to make sense, of course, the individuals in D have to be ordered: we could impose an ordering on any domain, but let's instead just assume that D is the set of natural numbers with their natural ordering. It is not essential to my argument that we use this particular relation nor that we order the domain. But a nice relation like this helps to keep the initial set up simple, and then one can generalise after seeing the key point. Notice that '⅄' is very similar to '∃' in that it is an existential quantifier, but it only 'looks' at a subset of the assignments that '∃' looks at. '∃' is the 'for some' quantifier, while '⅄' is the 'for some greater' quantifier.

Consider the following sentence of our language:

(9) ⅄xFx

Certainly, in (9) the variable in the embedded formula '*Fx*' is bound by the quantifier '⅄x'.

---

[5]Here we will not worry about the distinction between 'satisfaction by a sequence' and 'truth'—of course, Tarski reserves *truth* for formulae that are satisfied by all sequences.

Claim 1: All the variables in 'ɐxFx' are bound

This claim seems innocent enough, but some may suspect some kind of trickery: some sleight of hand with the operative notion of 'binding' or some variable up my sleeve. Are all the variables in 'ɐxFx' really bound? Standardly, an occurrence of a variable α is said to be *bound* in a formula just in case it is immediately attached to the quantifier or within the scope of a quantifier that is indexed with α (and free otherwise). So in a formula such as '(∃xFx ∧ Gy)', both occurrences of 'x' are bound, while the occurrence of 'y' is free. Clearly, given this standard definition, all the occurrences of variables in 'ɐxFx' are bound.

Thus if we can establish that prefixing a quantifier, such as '∃x', to (9) is not idle, then we will have a counterexample to the Principle of Vacuous Quantification. More precisely, we will demonstrate the following (for some model $\mathfrak{A}$):

Claim 2: $[\![∃xɐxFx]\!] \neq [\![ɐxFx]\!]$

We are assuming that D is the set of natural numbers, and let's also assume a particular interpretation for 'F', namely $I(F) = \{8\}$. So 'F' is only true of 8. Provided this model, consider the truth conditions of (9) relative to some assignment g, which are calculated as follows:

$[\![∃xɐxFx]\!]^g = 1$ iff for some g' that differs from g at most in that
$\qquad\qquad g'(x) > g(x)$, $[\![Fx]\!]^{g'} = 1$
$\qquad\qquad$ iff for some g' that differs from g at most in that
$\qquad\qquad g'(x) > g(x)$, $g'(x) \in I(F)$
$\qquad\qquad$ iff for some $n \geq g(x)$, $n = 8$.

Assume $g(x) = 10$, then since, of course, there isn't an $n \geq 10$ such that $n = 8$, it follows that $[\![ɐxFx]\!]^g = 0$. Now let's drop the hammer: embed 'ɐxFx' under '∃x':

(10) ∃xɐxFx

By calculating the truth conditions (relative to the same model), we see that the outer quantifier is not vacuous:

$[\![∃x⅄xFx]\!]^g = 1$ iff for some g' that differs from g at most in that
$g'(x) \neq g(x)$, $[\![⅄xFx]\!]^{g'} = 1$;
iff for some g' that differs from g at most in that
$g'(x) \neq g(x)$,
for some g'' that differs from g' at most in that
$g''(x) > g'(x)$, $g''(x) \in I(F)$;
iff for some m, for some $n \geq m$, $n = 8$.

Since there is an m and an n such that $n \geq m$ and $n = 8$, it follows that $[\![∃x⅄xFx]\!]^g = 1$. Thus, relative to this model and assignment g, $[\![⅄xFx]\!]^g = 0$, while $[\![∃x⅄xFx]\!]^g = 1$. This completes the proof that $[\![∃x⅄xFx]\!] \neq [\![⅄xFx]\!]$, and thus that the following biconditional is not valid (it is false at g):

(11) ∃x⅄xFx ↔ ⅄xFx

This provides a simple counterexample to the Principle of Vacuous Quantification. Thus even when all the variables in a formula ϕ are bound, prefixing a quantifier Σ to ϕ can be non-vacuous.[6]

Now let's take stock. If we restrict our focus to standard first-order logic, then the Principle of Vacuous Binding clearly holds, so what is the essential difference introduced by the '⅄' quantifier? First-order quantifiers (e.g. '∃' and '∀') are standardly appointed as special kind of stability—one that is not essential to their status as variable-binding operators. The set of a-variants that '∃' looks to when assessing its embedded formula are not at all constrained by what the initial variable assignment assigns to a, so shifting what the input assignment assigns to a will be idle. Whereas, the set of a-

---

[6]But what about the claim that bound variables can be re-bound? Does prefixing '∃x' to '⅄xFx' result in 'x' being rebound? I think so, but I'm not so interested in explicitly defending it here, since this turns on some subtle terminological issues. We'd need a definition of when a quantifier *binds* a particular occurrence of a variable in a formula. There is a standard definition of this in terms of syntax which stipulates that a bound variable can't be re-bound (Heim and Kratzer 1998, 120). So, even though a natural way to describe the counterexample above would be to say that '∃x' rebinds the last occurrence of 'x', this is ruled out by a standard definition. What's going on? Under the stress of the counterexample the syntactic definition of binding is pulling apart from the background semantic understanding of binding. The class of 'variables' and 'quantifiers' (or variables–binders in general) are grouped together due to their interesting *semantic* properties, not their syntactic properties. Thus in this more fundamental semantic sense, we should say that a quantifier binds an occurrence of a variable in a formula when the sensitivity of the variable is affected by the shifting induced by the quantifier. It is in the semantic sense that you can bind a bound variable.

variants that 'Ɐ' looks to when assessing its embedded formula *are* constrained by what the initial variable assignment assigns to *a*—they have to assign something greater than (or equal to) what the initial variable assignment assigns to *a*—thus shifting the input assignment can make a difference. In this way, '∃' standardly has a certain indifference to the input assignment, whereas for 'Ɐ' the input assignment genuinely matters—'Ɐ' is 'context' sensitive.[7]

This suggests that the essential difference between '∃' and 'Ɐ' concerns the *accessibility relations* involved. Thus it can be illuminating at this point to view first-order logic as a modal logic in the way associated with Amsterdam (see Van Benthem 1977 and especially Blackburn, De Rijke, and Venema 2002, Section 7.5 on reverse correspondence theory). On this way of viewing things, the assignments are the 'worlds', and the model includes a stock of binary accessibility relations R^α that hold between assignments (relative to a variable *a*). Standard first-order logic is only concerned with a special subset of all the possible such models for first-order languages—that is, it constrains itself to a particular accessibility relation:

$g \equiv^\alpha g'$ iff g' differs from g at most in that $g'(\alpha) \neq g(\alpha)$

Note that $\equiv^\alpha$ is reflexive, transitive, and symmetric. So given standard assumptions, the accessibility relation between assignments is an equivalence relation, and thus will validate the relevant formulae corresponding to S5, which will include the following theorems concerning vacuous quantification:

$\forall x \forall x \phi \leftrightarrow \forall x \phi$
$\exists x \forall x \phi \leftrightarrow \forall x \phi$
$\exists x \exists x \phi \leftrightarrow \exists x \phi$
$\forall x \exists x \phi \leftrightarrow \exists x \phi$

Prefixing further quantifiers to a closed sentence in standard first-order logic is 'vacuous'—just as adding further boxes and diamonds is vacuous in an S5 modal logic. But—just as in modal logic—the relevant equivalencies only hold given particular restrictions on the accessibility

---

[7] An alternative way to put the difference here is to say that standard first-order quantifiers take an 'external' perspective, whereas a quantifier like 'Ɐ' must take an 'internal' perspective on the relevant relational structures (see Recanati 2007, 65–71 and Blackburn, De Rijke, and Venema 2002, xi–x).

relation. For example, $\Box\phi \to \Box\Box\phi$ is invalid unless the frame is transitive. Likewise, if R^x is not transitive, then $\forall x\phi \to \forall x\forall x\phi$ will be invalid.

To round off this point, consider again the quantifier 'Я' and the accessibility relation that it appeals to:

$g \geq^\alpha g'$ iff g' differs from g at most in that $g'(\alpha) > g(\alpha)$

This relation is reflexive and transitive, but it is not symmetric. And our counterexample to the vacuous quantification principle implicitly exploited the fact that the relation was not symmetric. Notice that, in general, if the accessibility relation between assignments is not assumed to be symmetric then the paradigm example of vacuous quantification mentioned at the outset becomes invalid.

(8) $\forall x \exists x F x \leftrightarrow \exists x F x$

Consider the right-to-left direction. If someone taller than me is happy, it doesn't follow that everyone taller than me is such that there is someone taller than them that is happy.[8] One can imagine that by generalising and playing around with the first-order accessibility relations, there are things of intrinsic interest to metalogic, e.g. concerning decidability (see discussion in Blackburn, De Rijke, and Venema 2002, 466–469), but the application we are concerned with is modelling epistemic contexts—and the key insight here is that assignment-shifting operators can non-trivially stack.

2. Variables in counterpart semantics

As I mentioned at the outset, the discussion above has illuminating connections to Lewis' counterpart semantics (Lewis 1968). The connection is that in counterpart semantics the falsity of the Principle of Vacuous Binding is presupposed. Let me explain.

Lewis provides a first-order translation for formulae of quantified modal logic such as the following (where 'W' is the property of being a world, 'I' is

---

[8]There is a tight connection here between 'Я' and the kinds of quantifiers used for knowledge representation by description logics, e.g. '∃R' (see Baader et al. 2003; Blackburn 2006): they both restrict to the set of individuals that bear a relation to the input individual. The sentences of description logic function much like the sentences of Prior's Egocentric logic such as 'Someone-more-perfect standing', which is true at an individual a iff there is an x such that x is more perfect than a and x is standing (see Prior 1968).

the relation that holds between an object z and a world y when z is *in* y, and 'C' is the counterpart relation):

$$\Box Fx \approx \forall y \forall z((Wy \wedge Izy \wedge Czx) \rightarrow Fz)$$

This says (roughly) that every counterpart of x, in any world, is an F. And such a translation generalises for any modalised open sentence $\ulcorner \Box \pi \alpha_1 \ldots \alpha_n \urcorner$. Lewis demonstrates how we can follow the translation procedure, and then provide the resulting first-order formulae with their standard first-order interpretation, thereby endowing the modal formulae with truth conditions.

But the intermediate translation is not required. We can instead directly provide a model-theoretic semantics for the language of quantified modal logic that corresponds to Lewis' translation rules (see Hazen 1979, Cresswell and Hughes 1996, 353–358, and especially Schwarz 2012). And this is where things get interesting. Since we are only concerned with the model-theoretic semantics—not with certain metaphysical commitments—we will follow the advice of Schwarz (2012), who says:

> … we should dissociate counterpart semantics from various Lewisian doctrines that are commonly lumped together under the heading of 'counterpart theory'. (Schwarz 2012, 9)

So, in particular, we are not here concerned with modal realism, world-bound agents (i.e. Postulate 2), or Lewis' specific commitments on the similarity relation and counterparthood.[9] We are instead concerned with model-theoretic *counterpart semantics*.

For our purposes, the interesting action is going to be with modalised sentences such as '$\Box Fx$' and binding into such formulae. First, think about how we might interpret a modalised formulae such as '$\Box Fx$'. This says, roughly, that every counterpart of x, in any world, is an F, so we need to evaluate the embedded open formula 'Fx' relative to every counterpart of the individual assigned to x at every accessible world. Thus if '$\Box Fx$' is evaluated at a variable assignment g and world w, then 'Fx' must be evaluated at assignments g' where $g'(x)$ in a world is a 'counterpart' of $g(x)$ in w.

---

[9]Postulate 2 stipulates that 'Nothing is in two worlds'. In footnote 2, Lewis (1968) entertains the idea of allowing identities across worlds by giving up Postulate 2, but later clarifies that he is firmly committed to Postulate 2: 'Footnote 2 has given some readers the impression that I regard Postulate 2 as a mere convention, and that we could just as truly say that some things are identical with their otherworldly counterparts after all. Not so. I was alluding to the possibility of a hybrid theory—a theory opposed to my own, a theory which I take to be false—according to which there are identities across worlds, but we use the counterpart relation anyway' (Lewis 1983, 46). We are setting aside Lewis' particular metaphysical commitments that lead him to oppose the hybrid theory.

In this way, the modal shifts the assignment and effectively binds all the variables in its scope. In general, the clause for $\Box$ will yield:

⌜$\Box\phi$⌝ is true relative to a world w and assignment g iff ϕ is true relative to all w-accessible worlds w' and assignments g' that assign to each free variable α in ϕ a counterpart $g'(\alpha)$ at w' of $g(\alpha)$ at w.

The model, then, should include a relation on world-assignment pairs, which encodes the *counterpart relations*.

The counterpart relation and the assignment meddling of the modals are the most interesting departure from the basic Kripke semantics. The rest of the semantics is essentially the same. But just to have a full model on display, let's fill out the rest. Let the language be modal predicate logic with identity, as defined above. Let a model $\mathfrak{A} = \langle W, R, D, I \rangle$, where W is a (non-empty) set of worlds, R is a binary relation on $D^{\mathbb{N}} \times W$ (an accessibility relation incorporating the counterpart relation), D is a (non-empty) set of individuals, and I is the interpretation, which relative to a world assigns the predicate sets of tuples of individuals drawn from D.[10] We can then provide the semantic clauses relative to a model $\mathfrak{A}$ as follows:

- $[\![\alpha]\!]^{g,w} = g(\alpha)$
- $[\![\pi\alpha_1 \ldots \alpha_n]\!]^{g,w} = 1$ iff $\langle [\![\alpha_1]\!]^{g,w}, \ldots, [\![\alpha_n]\!]^{g,w} \rangle \in I(\pi, w)$
- $[\![\alpha = \beta]\!]^{g,w} = 1$ iff $[\![\alpha]\!]^{g,w} = [\![\beta]\!]^{g,w}$
- $[\![\neg\phi]\!]^{g,w} = 1$ iff $[\![\phi]\!]^{g,w} = 0$
- $[\![\phi \wedge \psi]\!]^{g,w} = 1$ iff $[\![\phi]\!]^{g,w} = 1$ and $[\![\psi]\!]^{g,w} = 1$
- $[\![\exists \alpha \phi]\!]^{g,w} = 1$ iff for some g' that differs from g at most in that $g'(\alpha) \neq g(\alpha)$, $[\![\phi]\!]^{g',w} = 1$
- $[\![\Box\phi]\!]^{g,w} = 1$ iff for all w' and g' such that $\langle w, g \rangle R \langle w', g' \rangle$, $[\![\phi]\!]^{g',w'} = 1$.

---

[10] Here I am loosely following Schwarz (2012), but there are a few differences in my presentation. Schwarz has the counterpart relation hold between individual-world pairs, and then uses this to construct the required alternative sequences (see p. 13ff), whereas I have a counterpart relation on sequence of individuals and world pairs. Both strategies allow for multiple counterparts at one world, but there are issues concerning multiply *de re* modality that I must gloss over here (see Hazen 1979, 328–330 and Lewis 1983, 44–45). I also assume a constant domain, while Schwarz doesn't. This is just for ease of presentation. The counterpart semantics can make all the same extra sophistications as the Kripke semantics can, e.g. with variable domains or inner/outer domains, etc. In fact, there are good reasons to add such sophistications. Here is one for the epistemic case: agents can think that there are more individuals than there in fact are. For example, assume the domain has two elements: {Pythagoras, Venus}. If Pythagoras can truthfully say 'Hesperus might not be Phosphorus', then there is a world where a Hesperus counterpart is distinct from a Phosphorus counterpart. But Pythagoras knows that he is distinct from both Hesperus and Phosphorus, so the only candidate for Hesperus- and Phosphorus-counterparts is Venus. Thus the 'might' claim isn't true unless we allow the domains to vary with worlds or we do some trick with inner/outer domains. See how to do it in Schwarz (2012).

Notice, again, in this last clause that the modal operator binds all the variables in its scope. But since the assignments that '$\Box$' looks to when assessing its embedded formula ϕ *are* constrained by what the initial variable assignment assigns to the variables in ϕ, prefixing a quantifier to '$\Box \phi$' needn't be vacuous. So although the modal operator binds the variables, they can be 're-bound'.

In fact, when addressing Kripke's 'purely formal objection' (1980, footnote 13), Lewis (1983)—the postscript to the 1968 article—explicitly points out that his modal operators bind the variables in their scope. The objection is that Lewis' theory invalidates certain cherished principles of the logic of identity and quantification. The way Lewis puts the objection is that his counterpart theory seems to invalidate Leibniz's law, $\ulcorner \forall x \forall y (x = y \rightarrow (\phi_x \leftrightarrow \phi_y)) \urcorner$, since the following is not valid:

(12) $\forall x \forall y (x = y \rightarrow (\Diamond x \neq y \leftrightarrow \Diamond y \neq y))$

But Lewis pleads 'not guilty'. He insists that on his view (12) is not really an instance of Leibniz's law for much the same reason that (13) isn't an instance.

(13) $\forall x \forall y (x = y \rightarrow (\exists y\ x \neq y \leftrightarrow \exists y\ y \neq y))$

Clearly, to think that (13) is an instance of Leibniz's law is, as Lewis says, 'to commit a fallacy of confusing bound variables'. Lewis insists that the same holds for (12), since the modals bind the variables in their scope.[11]

> The abbreviated notation of quantified modal logic conceals the true pattern of binding ... The diamonds conceal quantifiers that bind the occurrences of 'x' and 'y' that follow. ... So counterpart theory is no threat to standard logic. It is only a threat to simplistic methods of keeping track of variable-binding and instancehood when we are dealing with the perversely abbreviated language of quantified modal logic. (Lewis 1983, 46)

Thus this formal complaint against counterpart theory does not hold up to scrutiny.

The counterpart framework remains controversial. Some think that it has various attractive metaphysical applications, e.g. it avoids the

---

[11] See (Schwarz 2012, 17–18) for discussion of the appropriate restrictions on substitution in a counterpart semantics.

problem of accidental intrinsics, it allows for paradox-resolving flexibility in the attribution of modal properties, and allows for contingent identities, etc. While others think it is based on a confusion or inherits various implausible metaphysical commitments, or falls victim to the 'Humphrey objection'. My purpose is not to weigh in on these ongoing debates over the counterpart framework. But I do want to highlight that most of this controversy concerns whether Kripke's framework or Lewis' counterpart framework provides the most plausible analysis of *metaphysical* modality. The debate, however, looks very different in the context of epistemic modality. As I outlined already Kripke's framework seems ill-suited for epistemic modality *de re*, but moreover the standard complaints against Lewis' framework lose much of their force in the epistemic setting.

Consider the Humprey objection. Kripke complains,

> ... if we say 'Humphrey might have won the election (if only he had done such-and-such)', we are not talking about something that might have happened to Humphrey but to someone else, a 'counterpart'. Probably, however, Humphrey could not care less whether someone else, no matter how much resembling him, would have been victorious in another possible world. (Kripke 1980, 45)

Even if one finds this appeal convincing in the case of metaphysical possibility, it seems to miss the mark when aimed at a counterpart-theoretic analysis of *epistemic* modality. Kripke himself reaches for counterpart-theoretic devices when explaining the epistemic sense in which it might have turned out that Hesperus wasn't Phosphorus.

> And so it's true that given the evidence that someone has antecedent to his empirical investigation, he can be placed in a sense in exactly the same situation, that is a qualitatively identical epistemic situation, and call two heavenly bodies 'Hesperus' and 'Phosphorus', without their being identical. So in that sense we can say that it might have turned out either way. (Kripke 1980, 103–104)

Thus the epistemic modal claim is true in virtue of 'counterparts' of Hesperus and Phosphorus which are distinct in other worlds, not in virtue of Hesperus and Phosphorus themselves being distinct in other worlds. With the counterpart framework in mind, one might read Kripke as putting forward the following suggestion: an epistemic use of 'Hesperus might not be Phosphorus' is true iff there is a world w compatible with the speaker's qualitative evidence where an epistemic counterpart of Hesperus in w is distinct from an epistemic counterpart of Phosphorus in w.[12] In fact, Kripke encourages this counterpart-theoretic construal in the following passages[13]:

> Here, then, the notion of 'counterpart' comes into its own. For it is not the table, but and epistemic 'counterpart', which was hewn from ice; not Hesperus–Phosphorus–Venus, but two distinct counterparts thereof, in two of the roles Venus actually plays (that of Evening Star and Morning Star), which are different ... if someone confuses the epistemological and metaphysical problems, he will be well on the way to the counterpart theory of Lewis and other have advocated. (Kripke 1971, footnote 15)

> ... I (or some conscious being) could have been *qualitatively in the same epistemic situation* that in fact obtains, I could have the same sensory evidence that I in fact have, about *a table* which was made of ice. The situation is thus akin to the one which inspired the counterpart theorists. (Kripke 1980, 333)

The intuition that a certain table might have turned out to be made of ice concerns 'epistemic counterparts' of the table, the intuition that Hesperus could have turned out to be distinct from Phosphorus concerns distinct counterparts of Venus. Of course, Kripke is alluding to counterpart theory in a denigrating way: Lewis' counterpart theory makes sense for metaphysical modality only when you confuse it with epistemic modality. But this also suggests a positive position: Lewis' counterpart framework is well-suited for an analysis of epistemic modality.

3. Variables in epistemic contexts

Recently theorists have been appealing to 'assignment-shifting operators' in treatments of certain natural language constructions such as attitude verbs and epistemic modals, see, e.g. Cumming (2008), Santorio (2012), and Ninan (2012), Ninan (forthcoming). These views all fall within the family of counterpart semantics, broadly construed, and share the feature that the modals or attitude verbs bind all the variables in their

---

[12]The following passage also strongly suggests this reading: 'So two things are true: first, that we do not know a priori that Hesperus is Phosphorus, and are in no position to find out the answer except empirically. Second, this is so because we could have evidence qualitatively indistinguishable from the evidence we have and determine the referents of the two names be positions of the two planets in the sky, without the planets being the same'. (Kripke 1980, 104)

[13]Although one can read Kripke as suggesting this counterpart treatment of epistemic modality, I'm not claiming that this is Kripke's considered view. At times Kripke seems to suggest instead that talk of epistemic modality always involves a kind of 'loose speak'. On this reading, an epistemic use of 'Hesperus might not be Phosphorus' is strictly speaking false, but there is a *rephrasal* of it such as 'There is a world compatible with the speaker's qualitative evidence where an epistemic counterpart of Hesperus is distinct from an epistemic counterpart of Phosphorus', which is true. See Bealer (2002, 81–83) for a nice discussion of this point.

scope. But there are some differences in terms of motivation and implementation that are worth highlighting.

Cumming (2008) is concerned to provide a semantic view where belief attributions such as 'Biron thinks that Hesperus is visible' and 'Biron thinks that Phosphorus is visible' can differ in truth value. Cumming argues that both the Millian and the descriptivists views are untenable, even though they each harbour a half-truth: (i) Millianism is correct in maintaining that the referent of a name is not sensitive to the world parameter, (ii) Descriptivism is correct that the referent of a name is shiftable in epistemic contexts. We need an alternative that accommodates both of these features.

Cumming insists on two innovations. The first is that names should be semantically represented as variables, so that particular uses of 'Hesperus is visible' and 'Phosphorus is visible' will be equivalent to something like (14) and (15) respectively:

(14) x is visible
(15) y is visible

If variables are treated in the standard way, then they retain their status as insensitive to the world parameter. But they are sensitive to the assignment parameter, and this allows for the second innovation. The second innovation is that attitude verbs quantify over alternative assignments in addition to worlds. In Cumming's terminology, attitude verbs operate on *open propositions*, which are true and false with respect to a world and assignment. The justification for quantification over alternative assignments is that doxastic possibilities for an agent encode both information about the world and about the reference relation:

> ... verbs that create hyperintensional contexts, like 'think', are treated as operators that simultaneously shift the world and assignment parameters.... This conforms to the intuition that the content of attitude ascriptions encapsulates *referential uncertainty*.
>
> My treatment of attitude verbs as operators that shift the assignment tallies with the reflection that attitude ascriptions can convey things about how an agent conceives of the reference relation (in addition to how they conceive of the world). (Cumming 2008, 550)

One needn't construe the epistemic possibilities involved in this overtly meta-linguistic way. Formally, it amounts to the same thing but its perhaps better to construe the uncertainty involved as the agent's uncertainty about which individual x is, instead of the agent's uncertainty about which individual 'x' refers to. In any case, a belief report will have truth conditions such as the following:

> 'Biron believes x is visible' is true iff for each assignment-world pair ⟨g, w⟩ in Biron's belief set, g maps 'x' onto an object that is visible at w.

The official way that Cumming implements this in the model theory is as follows:

$[\![\Box_\tau \phi]\!]^{g,w} = 1$ iff for all g' and w' such that $\text{DOX}_\tau(w, \langle g', w'\rangle)$, $[\![\phi]\!]^{g',w'} = 1$,
where '$\Box_\tau$' is an agent relativised belief operator and doxastic accessibility for each agent τ is given by a relation DOX_τ between W and $D^{\mathbb{N}} \times W$.

On this approach, the truth of '$\Box_\tau Fx$' does not depend on what the input assignment assigns to 'x', thus we get the result that straightforward quantifying in with objectual binders is blocked. That is, for Cumming '$\Box$' not only unselectively binds all the variables in its scope it does so in a way that renders the resulting formula insensitive to the input assignment. Seemingly aware of this, Cumming (2008) instead appeals to substitutional quantification as a means of quantifying into attitude contexts—the machinery employed is inspired by Kaplan's (1968) treatment of *de re* attitude ascriptions. But Pickel (2015) has shown that the insensitivity of '$\Box_\tau Fx$' to the input assignment combined with the substitutional quantifiers nevertheless leads to undesirable results (see Pickel 2015, 339–341 and Rieppel 2017, 248–250).[14]

Santorio (2012) points to close cousins of the problems with names in attitude reports, namely cases where indexicals occurring under epistemic modals seem to shift. Consider the case of mad Heimson.[15]

> Heimson is a bit crazy and takes himself to be a philosopher of the Scottish Enlightenment, but he's uncertain which one he is: Stewart, Hume, or Smith. Alone in his study, he says to himself, 'I might be Hume'.

Since for all Heimson knows he is Hume, it seems that Heimson's utterance of (16) is true (or at least, true relative to his epistemic state).

---

[14]Pickel instead provides a two-factor view where the truth of '$\Box Fx$' depends on both (i) the truth of 'Fx' relative to shifted assignments (at the shifted worlds) and (ii) relative to the *input* assignment (at the shifted worlds). The latter conjunct gives rise to the arguably bad result that utterances of 'Olivia believes that Hesperus is distinct from Phosphorus' (and presumably 'Hesperus might not be Phosphorus') cannot be true—attitude reports have become *overly* sensitive to the input assignment. In terms of epistemic counterparts, Pickel's strategy allows for some of x's counterparts in other worlds to be distinct from x, but nevertheless requires that x itself is always among x's counterparts in other worlds.

[15]This is a somewhat simplified version of the kinds of examples that Santorio (2012) actually uses.

(16) I might be Hume

But on a standard treatment of indexicals and modals, Heimson's utterance would be false. The pronoun gets its value from the contextually determined assignment function—and thus refers to Heimson—while the modal quantifies over worlds compatible with Heimson's evidence. Thus Heimson's utterance of (16) is true only if there is a world where Heimson is identical to Hume. But there is no such world, so on the standard view (16) is false. Santorio makes the following suggestion:

> When 'I' occurs under an informational modal, it refers not to the actual speaker, but rather to representatives of the actual speaker in the relevant information state. Think of an information state as a set of possible worlds, namely, the worlds that are compatible with the relevant attitude.... Roughly, [the] representatives are individuals that, for all the subject knows, the speaker might be. (Santorio 2012, 373)

This is formally carried out by making a certain adjustment to the semantics of modals.

> On the standard view, informational modals are, in essence, quantifiers over possible worlds. On the view I'm advocating, they also encode in their meaning an apparatus that locates real-world individuals within the set of worlds quantified over. Thus on the new picture, these modals manipulate a greater amount of information ... [This] is implemented by letting epistemic operators manipulate the assignment parameter ... (Santorio 2012, 376)

Santorio builds contextually variable *counterpart functions* into the semantics, which encode a way in which the subject of the epistemic state is acquainted with various individuals. Formally, a counterpart function f is a function from worlds to individuals (i.e. an individual concept). For each variable $x_1, x_2, \ldots$, the context supplies both a value $g(1), g(2), \ldots$ for the variable and a counterpart function $f_1, f_2, \ldots$ associated with the variable. Then relative to an assignment g, a world w, and a sequence of counterpart functions $F = f_1, f_2, \ldots$, the clause for epistemic *might* and *must* is as follows:

$[\![\text{might } \phi]\!]^{g,w,F} = 1$ iff

(i) $[f_1(w) = g(1)] \wedge [f_2(w) = g(2)] \wedge [f_3(w) = g(3)] \wedge \ldots$
(ii) for some $\langle g', w' \rangle$ such that $w'Rw$ and $g' = \langle f_1(w'), f_2(w'), \ldots \rangle$, $[\![\phi]\!]^{g',w'} = 1$,

$[\![\text{must } \phi]\!]^{g,w,F} = 1$ iff

(i) $[f_1(w) = g(1)] \wedge [f_2(w) = g(2)] \wedge [f_3(w) = g(3)] \wedge \ldots$
(ii) for all $\langle g', w' \rangle$ such that $w'Rw$ and $g' = \langle f_1(w'), f_2(w'), \ldots \rangle$, $[\![\phi]\!]^{g',w'} = 1$.

Thus, (16), which is regimented as '◇ $x = h$' can come out true in a context.[16] It is true just in case the counterpart function f associated with 'x' picks out Heimson in the actual world and there is a world w' compatible with Heimson's information state such that 'x ' h' is satisfied by w' and a shifted assignment g', where $g'(x) = f(w')$—that is, just in case there is a world compatible with Heimson's information state where there is an 'epistemic counterpart' of Heimson who is Hume.[17]

Santorio's picture shares with Cumming's the feature that the epistemic operators unselectively bind all the variables in their scope. Santorio (2012) doesn't mention quantifying in, but prefixing an objectual quantifier to an epistemically modalised sentence won't end up being vacuous on his account. This is due to the 'check' on the counterpart functions in the first conjunct, which ensures that the counterpart function applied to the input world aligns with the initial assignment. Given this check, a modal sentence such as 'might ϕ' is still sensitive in the requisite way to the initial variable assignment.

Dilip Ninan has a series of papers (Ninan 2012, 2013, forthcoming), where he makes similar use of assignment-sensitive content and 'multi-centred worlds'. Ninan (2012) closely resembles the counterpart semantics described in the previous section, though it has additional bells and whistles.[18] In a more recent paper, Ninan (forthcoming) presents a puzzle, which can be used to motivate epistemic counterpart semantics.[19] The puzzle begins with this scenario:

> There is a lottery with only two tickets, a blue ticket and a red ticket. The tickets are also numbered 1 and 2, but we don't know which colour goes with which number. (Assume the number of the ticket is printed on one side of the ticket and the colour on the back.) The winner has been drawn, and we know that

---

[16] Here we treat 'Hume' as a constant h, but this is just for ease of illustration. On a fuller account, all names, pronouns, and variables will be rendered are assignment-sensitive terms.

[17] On this view, like the ones discussed above, an identity '$x=y$' will not be epistemically necessary, since the necessity claim will fail as long as there is an accessible point $\langle g, w \rangle$, where $g(x) \neq g(y)$.

[18] An earlier version of this paper surveyed Ninan (2012) and compared it to the views of Cumming and Santorio, but when doing the final revision for this paper Ninan sent me his new paper 'Quantification and Epistemic Modality', which I focus on here since it provides a strong argument for the epistemic counterpart semantics I outlined above.

[19] Closely related puzzles are discussed in Aloni (2001), Aloni (2005), and Gerbrandy (2000). Of course, epistemic puzzles in this general vicinity have been discussed for years under the heading of puzzles of *de re* beliefs.

the blue ticket won. But since we don't know whether the blue ticket is ticket 1 or ticket 2, we don't know the number of the winning ticket.

We can present the front and back of the tickets, in no particular order, as follows (with the blue ticket indicated as the winner):

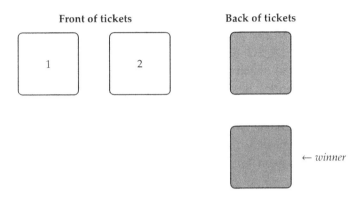

Ninan's puzzle proceeds by the following reasoning about this scenario:

(A) Ticket 1 is such that it might be the winning ticket.
(B) Ticket 2 is such that it might be the winning ticket.

Since those are the only tickets, it follows that any ticket is such that it might be the winning ticket. But the red ticket is a ticket, so it follows that:

(C) The red ticket is such that it might be the winning ticket.

Thus the puzzle is that from the apparently acceptable (A) and (B), we seem to be able to conclude the apparently unacceptable (C). (A) and (B) seem acceptable since we don't know the number of the winning ticket. But (C) seems unacceptable since we know that the red ticket lost. The reasoning from (A) and (B) to (C) seems valid, and given the standard Kripke-style semantics, it is valid. As Ninan notes the puzzle specifically concerns *epistemic* modals—there is no analogous puzzle with metaphysical modals (e.g. 'The red ticket might have been the winning ticket' is perfectly acceptable). As an account of epistemic modality, the standard view is faced with the choice of denying the conjunction of (A) and (B) or accepting (C). Neither option looks plausible.

This provides motivation for seeking an alternative account—of course, it is more-or-less the same kind motivation I've been harping on at various points in this paper, but Ninan's puzzle affords a clear and concise way of making the problem vivid. Ninan outlines two alternatives, which solve the puzzle. Importantly both alternatives abandon the 'objectual' aspect of standard Kripke semantics, whereby variables are both rigid and immune to shifting by modals. Ninan insists that the way forward is to take on board the *Quinean insight* that 'being necessarily or possibly thus and so is not a trait of the object concerned, but depends on the manner of referring to the object' (Quine 1953a, 148).

The first alternative Ninan develops builds on the dynamic proposal of Yalcin (2015) by supplementing it with Carnapian individual concepts (in terms of *conceptual covers*, see Aloni 2001), while the second alternative generalises the static semantics with counterpart relations. The key to both approaches is that they allow us to say that object o' in world w' represents object o in world w even if o is not identical to o'. To implement the counterpart version of this, Ninan adopts the kinds of models we defined above, where $\mathfrak{A} = \langle W, R, D, I \rangle$. And then introduces counterpart relations which are binary relations on $D \times W$. Second, he construes the counterpart relation as a parameter of the index (instead of as fixed in the model).[20] Thus he defines the truth of a formula relative to a counterpart relation K in addition to a world w and a variable assignment g (and a model $\mathfrak{A}$). All clauses ignore the counterpart parameter and are thus essentially the same as on the Kripke-semantics, save for epistemic 'must' and 'might', which are defined as follows:

- $[\![\Box \phi]\!]^{g,w,K} = 1$ iff $[\![\phi]\!]^{g',w',K} = 1$,
for all w' such that wRw' and for all g' such that for each variable i, $\langle g(i), w \rangle K \langle g'(i), w' \rangle$.
- $[\![\Diamond \phi]\!]^{g,w,K} = 1$ iff $[\![\phi]\!]^{g',w',K} = 1$,
for some w' such that wRw' and for some and g' such that for each variable i, $\langle g(i), w \rangle K \langle g'(i), w' \rangle$.

We can now model the lottery scenario. Let 'W' be the predicate for 'winning ticket', let 't_1' go proxy for 'ticket 1', and let 't_2' go proxy

---

[20] Schwarz (2012) and Cresswell and Hughes (1996) provide essentially the same semantics as Ninan, but they define the counterpart relation as an element of the model. The difference here doesn't matter much in terms of the pure semantics but may lead to differences in implementation with the post-semantics definition of truth, under the assumption that context supplies the counterpart relation. Presumably Ninan is anticipating a standard Kaplanian definition of truth-in-a-context, and he includes the counterpart relation as a parameter of the index in order to secure the contextual variability of the counterpart relation. That is, it is an important aspect of this approach that which object-in-a-world represents another object-in-a-world is something that can vary with the utterance context.

for 'ticket 2'. Then (A) and (B) will be rendered in our language as follows[21]:

(A') $\Diamond Wt_1$
(B') $\Diamond Wt_2$.

The truth conditions for each of these are as follows:

A': $[\![\Diamond Wt_1]\!]^{g,w,K} = 1$ iff $[\![Wt_1]\!]^{g',w',K} = 1$,

for some w' such that wRw' and for some and g' such that $\langle g(t_1), w \rangle K \langle g'(t_1), w' \rangle$.

B': $[\![\Diamond Wt_2]\!]^{g,w,K} = 1$ iff $[\![Wt_2]\!]^{g',w',K} = 1$,

for some w' such that wRw' and for some and g' such that $\langle g(t_2), w \rangle K \langle g'(t_2), w' \rangle$.

To make things concrete, assume the domain only has two tickets so that $D = \{a,b\}$ and assume that $g(t_1) = a$ and $g(t_2) = b$. Then we can see that (A') is true just in case there is an accessible world where a counterpart of a under relation K is W, and (B') is true just in case there is an accessible world where a counterpart of b under relation K is W. Thus the truth of the conjunction of (A') and (B') entails that anything might be W:
$[\![\forall x \Diamond Wx]\!]^{g,w,K} = 1$
 iff for all g' that differ from g at most in that $g'(x) \neq g(x)$, $[\![Wx]\!]^{g'',w',K} = 1$,

for some w' such that wRw' and for some and g'' such that $\langle g'(x), w \rangle K \langle g''(x), w' \rangle$.

That is, anything might be W iff for each individual x in the domain there is an accessible world where a counterpart of x under K is W. And that follows if:

(i) a and b are the only members of the domain;

---

[21] Ninan symbolises these with lambda binding—$(\lambda x. \Diamond Wx)(t_1)$ and $(\lambda x. \Diamond Wx)(t_2)$—in order to ensure that the modal predications are read *de re*. After all the stilted 'is such that' locution is employed so that the sentences are read *de re*. We could easily add lambdas to our language to do this as well, or we could fake it with constructions such as '$t_1 = x \wedge \Diamond Wx$'. But we needn't do this—Ninan does because he is being neutral on whether or not t_1 and t_2 are descriptions or names. Since we are assuming that t_1 and t_2 are variables, these further complications are actually unnecessary.

(ii) there is an accessible world where a counterpart of a under K is W;
(iii) there is an accessible world where a counterpart of b under K is W.

But now what about the crucial sentence (C): 'The red ticket is such that it might be the winner'. Let 'r' go proxy for 'The red ticket' (where again we are reading the sentence *de re* so we are assuming that 'r' is a name or variable.) So we have

(C') $\Diamond Wr$

And its truth conditions are given as follows:
C': $[\![\Diamond Wr]\!]^{g,w,K} = 1$ iff $[\![Wr]\!]^{g',w',K} = 1$,

for some w' such that wRw' and for some and g' such that $\langle g(r), w \rangle K \langle g'(r), w' \rangle$.

Given that a and b are the only things in the domain, $g(r) = a$ or $g(r) = b$. And on either choice, we can see that (C') follows from (A') and (B'). That is, (C') is true just in case there is an accessible world where a counterpart of $g(r)$ under K is W. And that is guaranteed given the truth of both (A') and (B') (plus the assumptions about the domain and that $g(t_1) \neq g(t_2)$).

At this point, one might think the counterpart theory has failed to diagnose the puzzle, since the argument comes out as valid! The whole problem was that we want to accept (A') and (B') without accepting (C'). So how is this any help? While the argument is formally valid according to the counterpart semantics, the story I've told so far hasn't made full use of the counterpart-theoretic resources. In giving the formal argument, we've just assumed some generic counterpart relation that we've held fixed throughout. Yet Ninan's lottery scenario is specifically designed to make two counterpart relations salient:

K^n = the *number counterpart relation*, which holds between ticket t in world w and ticket t' in world w' iff t in w and t' in w' have the same number.
K^c = the *colour counterpart relation*, which holds between ticket t in world w and ticket t' in world w' iff t in w and t' in w' are the same colour.

The counterpart-theoretic diagnosis of the puzzle insists that we want to accept (A') and (B') relative to counterpart relation K^n and we want to deny (C') relative to counterpart relation K^c. Although (C') is true relative to K^n, it is practically impossible to utter 'The red ticket is such that it might be the winner' without making the colour counterpart relation salient. Likewise although (A') and (B') are false relative to K^c it is difficult, if not impossible, for that counterpart relation to be salient in a context in which *a* and *b* are referred to as *Ticket 1* and *Ticket 2*. Thus the solution in terms of epistemic counterpart semantics is a contextualist solution—there is contextual variability of the counterpart relation (cf. Lewis 1983, 43–43; Lewis 1986, 251–263), and the contexts in which we want to accept the premises are different from the contexts in which we want to deny the conclusion.

This proposal is similar in spirit to various proposals that have been around forever. In particular, epistemic counterpart semantics bears an affinity to both a Carnapian semantics in terms of individual concepts (Carnap 1947; Aloni 2005) and epistemic two-dimensionalism (Chalmers 2004). In fact, some of the formal devices that are employed by these approaches are more-or-less the same. There are many choice points within these frameworks and given the right tweaks on both ends, we might even get to the same mathematical structure.[22]

Even though the picture shares some features with descriptivist views, it doesn't seem threatened by familiar anti-descriptivist arguments. It is important to notice that the epistemic counterpart relations that are associated with terms are not lexicalised, they are instead determined by the context. (For example, the number relation and the colour relation are not somehow lexically attached to 'Ticket 1' and 'The red ticket', respectively.) Thus there is no commitment to the idea that names (or variables) are synonymous with descriptions, nor to the idea that the referent of a name is determined via description. Kripke's (1980) arguments, then, can't seem to get off the ground, if aimed at the epistemic counterpart approach. For example, consider the argument from ignorance. If in a given context *c*, the subject thinks of Cicero simply as 'a Roman orator', then 'Cicero' in that context still manages to refer to Cicero because the way the subject thinks of Cicero is irrelevant to the determination of the referent of 'Cicero'. Of course, the way the subject thinks of Cicero, in a sense, determines the referent of 'Cicero' in the agent's 'epistemic worlds'—in every world compatible with the subject's knowledge the epistemic counterparts of Cicero are Roman orators. But it is not clear what the objection to this could be. Whereas with standard descriptivism, Carnapian

---

[22]One obvious difference is that individual concepts and primary intensions are functional, whereas counterpart relations are not so constrained. Ninan's (forthcoming) *one hundred ticket* case brings out this difference.

individual concepts or epistemic two-dimensionalism the function from worlds to individuals is lexically associated with the singular term, and it is this assumption in particular that is leveraged against those views. Of course, such views could be modified so that the mode of presentation is detached from reference determination and so that there is contextual variability of the relevant intensions (see Aloni 2005). There has yet to be a detailed survey of these various accounts comparing them formally, and mapping out where there are genuine differences and where the differences are merely notational. This should be done. But in any case, I predict that the flexibility and generality of the epistemic counterpart approach outlined above will be seen as a virtue.

Acknowledgements

For helpful comments and feedback, many thanks to Wolfgang Schwarz, Landon Rabern, Dilip Ninan, Paolo Santorio, Martin Smith, Anders Schoubye, David Chalmers, Bryan Pickel, and an anonymous referee, as well as audiences at the MCMP Colloquium in Munich, Toronto's M&E Group, the St Andrews Philosophy Club, the Edinburgh Language Workshop, and the Barcelona Workshop on Operators vs Quantifiers.

Disclosure statement

No potential conflict of interest was reported by the author.

References

Aloni, Maria. 2001. *Quantification under Conceptual Covers*. Phd. thesis. Amsterdam: University of Amsterdam.
Aloni, M. 2005. "Individual Concepts in Modal Predicate Logic." *Journal of Philosophical Logic* 34 (1): 1–64.
Baader, F., D. Calvanese, D. McGuinness, D. Nardi, and P. Patel-Schneider. 2003. *The Description Logic Handbook: Theory, Implementation and Applications*. New York: Cambridge University Press.
Barcan, R. C. 1946. "A Functional Calculus of First Order Based on Strict Implication." *The Journal of Symbolic Logic* 11 (1): 1–16.
Bealer, G. 2002. "Modal Epistemology and the Rationalist Renaissance." In *Conceivability and Possibility*, edited by T. S. Gendler and J. Hawthorne, 71–125. Oxford: Oxford University Press.
Blackburn, P. 2006. "Arthur Prior and Hybrid Logic." *Synthese* 150 (3): 329–372.
Blackburn, P., M. De Rijke, and Y. Venema. 2002. *Modal Logic*. Vol. 53. Cambridge: Cambridge University Press.
Burgess, J. P. 1998. "Quinus ab omni naevo vindicatus." In *Meaning and Reference*, edited by A. A. Kazmi, 25–66. Calgary: University of Calgary Press.

Carnap, R. 1947. *Meaning and Necessity: A Study in Semantics and Modal Logic.* Chicago: University of Chicago Press.
Chalmers, D. J. 2004. "Epistemic Two-Dimensional Semantics." *Philosophical Studies* 118 (1): 153–226.
Chalmers, D. J. 2011. "The Nature of Epistemic Space." In *Epistemic Modality*, edited by A. Egan and B. Weatherson. Oxford: Oxford University Press.
Chalmers, D. J., and B. Rabern. 2014. "Two-Dimensional Semantics and the Nesting Problem." *Analysis* 74 (2): 210–224.
Chalmers, D. J. unpublished. "The Tyranny of the Subjunctive." consc.net/papers/tyranny.html.
Cresswell, M. J., and G. E. Hughes. 1996. *A New Introduction to Modal Logic.* New York: Routledge.
Cumming, S. 2008. "Variablism." *Philosophical Review* 117 (4): 605–631.
Davies, M., and L. Humberstone. 1980. "Two Notions of Necessity." *Philosophical Studies* 38 (1): 1–30.
Egan, A., and B. Weatherson. 2011. *Epistemic Modality.* Oxford: Oxford University Press.
Gerbrandy, J. 1997. "Questions of Identity." In *Proceedings of the Eleventh Amsterdam Colloquium*, edited by P. Dekker, M. Stokhof and Y. Venema. Amsterdam: ILLC, University of Amsterdam.
Gerbrandy, J. 2000. "Identity in Epistemic Semantics." In *Logic, Language and Computation, Vol. III*, edited by P. Blackburn and J. Seligman. Stanford, CA: CSLI.
Hazen, A. 1979. "Counterpart-Theoretic Semantics for Modal Logic." *The Journal of Philosophy* 76 (6): 319–338.
Heim, I., and A. Kratzer. 1998. *Semantics in Generative Grammar.* Oxford: Wiley-Blackwell.
Kalish, D., and R. Montague. 1964. *Logic: Techniques of Formal Reasoning.* Oxford: Oxford University Press.
Kaplan, D. 1968. "Quantifying in." *Synthese* 19 (1–2): 178–214.
Kratzer, A. 1977. "What 'Must' and 'Can' Must and Can Mean." *Linguistics and Philosophy* 1 (3): 337–355.
Kripke, S. 1963. "Semantical Considerations on Modal Logic." *Acta Philosophica Fennica* 16: 83–94.
Kripke, S. 1971. "Identity and necessity." In *Identity and Individuation*, edited by M. K. Munitz, 135–164. New York City: New York University Press.
Kripke, S. 1980. *Naming and Necessity.* Cambridge, MA: Harvard University Press.
Lennertz, B. 2015. "Quantificational Credences." *Philosophers' Imprint* 15 (9): 1–24.
Lewis, D. K. 1968. "Counterpart Theory and Quantified Modal Logic." *The Journal of Philosophy* 65:113–126.
Lewis, D. 1983. "Postscripts to 'Counterpart Theory and Quantified Modal Logic'." In *Philosophical Papers I*, 39–46. Oxford: Oxford University Press.
Lewis, D. K. 1986. *On the Plurality of Worlds.* Oxford: Blackwell.
Ninan, D. 2012. "Counterfactual Attitudes and Multi-Centered Worlds." *Semantics and Pragmatics* 5 (5): 1–57.
Ninan, D. 2013. "Self-Location and Other-Location." *Philosophy and Phenomenological Research* 87 (1): 301–331.
Ninan, D. forthcoming. "Quantification and Epistemic Modality." *Philosophical Review.*
Pickel, B. 2015. "Variables and Attitudes." *Noûs* 49 (2): 333–356.

Prior, A. 1968. "Egocentric Logic." *Noûs* 2 (3): 191–207.
Quine, W. V. 1947. "The Problem of Interpreting Modal Logic." *The Journal of Symbolic Logic* 12 (2): 43–48.
Quine, W. V. 1953a. "Reference and Modality." In *From a Logical Point of View*, 139–159. Cambridge: Harvard University Press.
Quine, W. V. 1953b. "Three Grades of Modal Involvement." Proceedings of the XIth International Congress of Philosophy, Brussels, Vol. 14, 65–81
Recanati, F. 2007. *Perspectival Thought: A Plea for (Moderate) Relativism*. Oxford: Oxford University Press.
Rieppel, M. 2017. "Names, Masks, and Double Vision." *Ergo* 4 (8): 229–257.
Santorio, P. 2012. "Reference and Monstrosity." *Philosophical Review* 121 (3): 359–406.
Schwarz, W. 2012. "How Things are Elsewhere: Adventures in Counterpart Semantics." In *New Waves in Philosophical Logic*, edited by G. Restall and G. Russell, 8–29. Basingstoke: Palgrave Macmillian.
Swanson, E. 2010. "On Scope Relations Between Quantifiers and Epistemic Modals." *Journal of Semantics* 27 (4): 529–540.
Van Benthem, J. 1977. "Correspondence Theory." PhD thesis, University of Amsterdam.
Von Fintel, K., and S. Iatridou. 2003. "Epistemic Containment." *Linguistic Inquiry* 34 (2): 173–198.
Yalcin, S. 2015. "Epistemic Modality *de re*." *Ergo* 2 (19): 475–527.

Appendix. Quantified metaphysical and epistemic modality

I present a simple language incorporating the lessons from above. The language will be a mixed modal language with both objective modals and epistemic modals, and it will also have individual quantifiers.[23] Interestingly, we can model some version of Kripke's cases of the a posteriori necessities and a priori contingencies, in way that (prima facie) differs from the two-dimensionalist models (e.g. Davies and Humberstone 1980). Here's the language.

Variables: $x, y, z, \ldots$
Predicates: $F, G, H, \ldots$
Constants: $\neg, \wedge, \exists, =, \Box, \blacksquare$

For any sequence of variables, $\alpha_1, \ldots, \alpha_n$, any *n*-place predicate π, and any variables α and β, the sentences of the language are provided by the following grammar:

$$\phi ::= \pi\alpha_1 \ldots \alpha_n \mid \alpha = \beta \mid \neg\phi \mid (\phi \wedge \phi) \mid \exists\alpha\phi \mid \Box\phi \mid \blacksquare\phi.$$

Let a model $\mathfrak{A} = \langle W, R, K, D, I \rangle$, where W is a (non-empty) set of worlds, R is a binary relation on W (a metaphysical accessibility relation), K is a binary relation on $D^{\mathbb{N}} \times W$ (an epistemic accessibility relation), D is a (non-empty) set of individuals, and I is the interpretation, which relative to a world assigns the predicate sets of tuples of

---

[23]It is not strictly necessary to lexically distinguish the metaphysical and epistemic modals, I do that here just for perspicuity. One could easily make this in line with the proposal of Kratzer (1977), by letting the context fix the accessibility relation between assignment-world pairs, where metaphysical contexts would turn out to set the counterpart relation to identity.

individuals drawn from D. (Again we present a constant domain semantics for ease, while we really prefer a variable domain.) We then provide the semantic clauses relative to a model $\mathfrak{A}$ as follows:

- $[\![\alpha]\!]^{g,w} = g(\alpha)$
- $[\![\pi\alpha_1 \ldots \alpha_n]\!]^{g,w} = 1$ iff $\langle [\![\alpha_1]\!]^{g,w}, \ldots, [\![\alpha_n]\!]^{g,w} \rangle \in I(\pi,w)$
- $[\![\alpha = \beta]\!]^{g,w} = 1$ iff $[\![\alpha]\!]^{g,w} = [\![\beta]\!]^{g,w}$
- $[\![\neg\phi]\!]^{g,w} = 1$ iff $[\![\phi]\!]^{g,w} = 0$
- $[\![\phi \wedge \psi]\!]^{g,w} = 1$ iff $[\![\phi]\!]^{g,w} = 1$ and $[\![\psi]\!]^{g,w} = 1$
- $[\![\exists\alpha\phi]\!]^{g,w} = 1$ iff for some g' that differs from g at most in that $g'(\alpha) \neq g(\alpha)$, $[\![\phi]\!]^{g',w} = 1$
- $[\![\Box\phi]\!]^{g,w} = 1$ iff for all w' such that wRw', $[\![\phi]\!]^{g,w'} = 1$
- $[\![\blacksquare\phi]\!]^{g,w} = 1$ iff for all w' and g' such that $\langle w,g\rangle K\langle w',g'\rangle$, $[\![\phi]\!]^{g',w'} = 1$.

This kind of model can deal with the puzzles concerning epistemic modality *de re*, that we have covered in the paper.[24] It also affords a kind of diagnosis of the Kripkean necessary a posteriori and contingent a priori. Here is a brief sketch. First, a posteriori necessities:

(17) Hesperus couldn't have failed to be Phosphorus
    ::$\Box\, h = p$
(18) Hesperus might not be Phosphorus
    ::$\blacksquare\neg\, h = p$

The metaphysical claim (17) is straightforwardly true, assuming h=p; while the epistemic *might*-claim (18) is also true under certain epistemic counterpart relations, e.g. the counterpart relations that would be relevant in a context in which Pythagorus utters it. Next, cases of the contingent a priori.

(19) Julius has to be the inventor of the zip
    ::$\blacksquare\, Zj$
(20) Julius could have failed to invent the zip
    ::$\neg\Box\, Zj$

Of course the truth of (19) depends on which counterpart relation is contextually salient. And it is plausible that in a context where we have just introduced the name 'Julius' as a name of the inventor of the zip, every epistemic counterpart of Julius will be a zip inventor. Nevertheless, (20) is true for the familiar reason that zip-inventing is a contingent property of Julius.[25]

---

[24] I have included the counterpart relation as a part of the model, but for the reasons mentioned in footnote 20, if this type of model is employed in a natural language setting where there is contextual variability we ought to construe the counterpart relation as an index in the point of reference.

[25] An open question is how this system fares in terms of what Chalmers and Rabern (2014) call *the generalised nesting problem*, although it appears promising since it would seem to invalidate (A2).

Operators vs. quantifiers: the view from linguistics

Ariel Cohen

ABSTRACT
In several publications, François Recanati argues that time, world, location, and similar constituents are not arguments of the verb, although they do affect truth conditions. However, he points out that this fact does not decide the debate regarding whether these notions are represented as sentential operators variables bound by quantifiers, as both approaches can be made compatible with such non-arguments. He makes these points using philosophical arguments; in this paper I use linguistic evidence from a variety of languages. Specifically, I demonstrate that time, world, location, and person behave syntactically and semantically the same across languages. Hence, either all are arguments, or neither are; and the evidence points to the latter position. The cross-linguistic data give rise to a generalization, which connects the availability of quantifying over times, worlds, locations, or persons with the availability of indexical expressions of the corresponding type. For example, a language that does not have temporal indexicals cannot have tense. I demonstrate that this generalization follows naturally from the quantifier approach, but would remain a mystery under the operator view. Therefore, the cross-linguistic evidence argues in the favor of the quantificational approach.

1. Introduction

How do we represent time in natural language? How can we express the fact that (1) is about a past event?

(1)   John sang.

Time can be represented in one of two formalizations. One possibility is using an intensional tense operator. In this case, the logical form of (1) would be something like (2).

(2)   P **sing(j)**

An alternative is an extensional framework, which explicitly expresses quantification over times (where t_0 is the time of evaluation):

(3) $\exists\, t < t_0$ **sing(j,t)**

The same debate holds regarding modality: is it expressed by an operator, or by explicit quantification over worlds? And, as we shall see, the debate can be applied to other notions as well; in particular, we will also discuss location and person.

How do we decide between the quantifier and the operator approaches? Here is one direction that seems promising: investigate whether time, world, etc. are arguments of the verb. It would seem that, if they are, this would support the quantificational view, whereas if they are not, this could be construed as evidence for the operators view.

Things are not so simple, however. In a series of publications, François Recanati (2002, 2007a, 2007b) argues forcefully for the view that such notions are not part of the predicate-argument structure.[1] But he also shows that even if his claim is granted, this would not decide the issue: both approaches can be made compatible with it.[2]

Is there no way out? Do the arguments have to be based solely on theoretical considerations, or can we find empirical arguments for or against one of the theories?

I believe that part of the problem is the fact that the debate over these issues has focused almost exclusively on English. But therein lurks a danger: linguistic analyses may be confused with conceptual ones, and idiosyncratic properties of English might be mistaken for deep, necessary principles.

This is where an empirical cross-linguistic investigation can be extremely effective. By looking at a variety of languages, from different language families, we can abstract over specific, idiosyncratic properties, and identify the underlying principles. This is the kind of approach that I am proposing in this paper. It should be noted that the purpose of the linguistic empirical observations and linguistic theoretical considerations is to provide an argument for a philosophical point: specifically, I will argue for the quantificational approach.

---

[1] Recanati argues for this position partly on the basis of *innocence*: the notion that a communicating agent may be unaware of a notion such as the passage of time or alternative possible worlds, and unable to think or speak about such a notion. This proposal, in fact, has linguistic implications that can be empirically tested; unfortunately, a discussion of innocence lies outside the scope of this paper.
[2] See Section 3 below.

The organization of this paper is as follows. In the next section, I will consider a number of notions: time, possible world, location and person. I will discuss the philosophical arguments for their status as non-arguments, point out the linguistic implications of these proposals and test them empirically. In Section 3, I discuss Recanati's proposal of *variadic functions*, whose role is to allow the quantificational approach to be compatible with the view that time, world, etc. are not arguments. I will formulate the logical forms for the notions discussed above, and demonstrate that all of them rely crucially on indexicals. In Section 4, I formulate the following principle: if a language does not have an indexical of a certain sort, the language cannot quantify over elements of that sort. For example, a language without temporal indexicals cannot have tense. I confirm this principle across a number of languages, and demonstrate that it follows from the quantifier, but not the operator, approach. Section 5 is a conclusion.

2. Philosophical theory and linguistic evidence

2.1. Time

Consider the following sentence, from Recanati (2002):
(4)   I've had a very large breakfast.

Normally, we interpret the time of the breakfast to be on the day of utterance. Recanati demonstrates that this interpretation is part of the semantics of the sentence, in the sense that it can serve as the input to implicature. For example, when uttered as a response to an invitation to have lunch, (4) clearly implicates that the speaker is not hungry; but this implicature can only come about if the breakfast is interpreted to take place at the same day as the utterance.

The time is clearly unarticulated, in the sense that there is no overt indication of the time of the breakfast in the sentence. But maybe it *is* articulated in the predicate-argument structure? Maybe tense provides a variable, which is an argument of the verb, to be filled by the context?[3] Recanati (2007a) answers in the negative. He argues that, in general, the event time is not an argument of the verb: 'While some verbs, like "last", have a temporal argument, others like "dance", do not' (135).

This intuition, although surprisingly hard to define precisely, is, in fact, shared by linguists. Grimshaw (1994) notes: '*last* has a temporal argument,

---

[3]Times are to be distinguished from events, which, according to some views, *are* arguments of the verb. For the distinction between times and events see, *inter alia*, Glasbey (1992) and Ramchand (2004).

while *wriggle* occurs with a temporal adjunct' (417). Thus, while (5a) can be paraphrased by (5b), (6a) cannot be paraphrased by the nonsensical (6b).

(5) a. The performer wriggled for an hour.
    b. The performer wriggled, and the duration of this event was one hour.
(6) a. The performance lasted for an hour.
    b. *The performance lasted, and the duration of this event was one hour.

What sort of linguistic evidence can we bring to bear on Recanati's position?

To indicate time, the verb is usually inflected for tense. In fact, Chomsky (1986) takes this inflection to be the defining characteristic of sentences; he goes so far as to replace the term *sentence* with Inflectional Phrase, or IP. Chomsky proposes that inflection is a linguistic category with a dedicated syntactic position – INFL – which is above the Verb Phrase (VP).

For example, consider the following simple sentence:

(7)   John likes his teacher.

Its structure would be represented as follows:

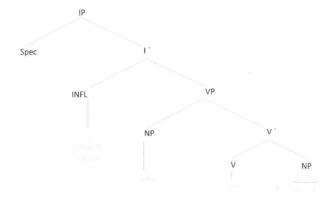

The node INFL contains the morpheme -*s*, which indicates the present tense.[4] Note that, according to current syntactic theory, the subject – *John* – is inside the VP (Koopman and Sportiche 1991).

---

[4]This morpheme also indicates third-person singular agreement. Indeed, Chomsky proposes that INFL represents agreement in addition to tense, but we will not deal with agreement here.

The standard linguistic assumption is that all and only material inside the VP is part of the predicate-argument structure. Since INFL is outside the VP, but all the arguments of the verb, including the subject, are inside, it follows that time is not a *syntactic* argument of the verb.

This, of course, does not yet answer the question of whether time is an argument. It might be argued, *contra* Recanati, that although time is not a syntactic argument of the verb, it is, nevertheless, a *semantic* argument[5]

The notions of syntactic argument and semantic argument are related, but quite distinct. An important discussion of this distinction is Chung and Ladusaw (2004). They choose Chamorro as a test case, and demonstrate that, in this language, a particular phenomenon, which they call *the extra object*, is a syntactic adjunct yet a semantic argument.

Since the linguistic representation of time resides in INFL, the question whether time is a semantic argument of the verb is now reduced to the question whether the material in INFL is a semantic argument of the verb. It might seem that this is merely restating the philosophical problem in technical linguistic terms; but this is not the case. By reducing the philosophical question to a linguistic one we *have* made some progress. The reason is that, as we will see shortly, recent linguistic evidence suggests that the material in INFL is not restricted to time only. INFL may contain location, possible world, or person. And INFL either is or is not a semantic argument of the verb, regardless of the material that is represented there.

This point merits some discussion. Crucially, whether a particular syntactic position is associated with a semantic function – being or not being an argument – obtains regardless of the content it contains. In particular, the content may be human, animate, inanimate, location, time, etc.; and either in all these cases this position is a semantic argument of the verb, or in all these cases this position is *not* a semantic argument of the verb.

To give the simplest of examples: the sister of the verb is usually considered to be its semantic argument. This holds regardless of whether this sister refers to an inanimate object, a human being, a location, a time, or what have you:

(8) a. The Boxer Hit the Boxing Bag
 b. The boxer hit his opponent.

---

[5] Of course, this does not preclude that it is an argument of higher elements. For example, the event argument is often thought to originate in a higher projection as an argument of the entire VP. But the important point is that it is not an argument of the verb.

c. The boxer arrived at the arena.
d. The boxer lasted for three minutes.

Indeed, this notion, that whether a particular syntactic position is or is not a semantic argument does not depend on its content, underlies the concept of *selectional restrictions*. Crucially, these are restrictions that a predicate places on its semantic arguments. Jackendoff (1992) defines them as follows:

> Selectional restrictions are general semantic restrictions on arguments ... Selectional restrictions evidently are constructed out of a subvocabulary of conceptual structures. That is, the set of possible selectional restrictions is chosen from primitives and principles of combination present in conceptual structure, including not only major conceptual category but also distinctions such as solid versus liquid, human versus animal, and so on. Thus the appropriate linguistic level for stating them is conceptual structure and not syntax or a putative level of argument structure. (51–52)

That is to say, selectional restrictions do not determine whether a linguistic expression in a particular position is an argument or not; the expression can denote a physical object, an abstract object, a time, a location, or whatever, and its argumenthood will be determined by syntax. Only at *that* point does the linguistic system access its semantic category, and then the sentence can be usefully interpreted or rejected as nonsensical.

From the discussion above it follows that either all the notions that are represented in INFL are arguments, or none of them are. Therefore, it would be enough to establish that *one* of them is, or is not, an argument, and the status of all the others would follow immediately. In other words, while it might be difficult to determine whether time is an argument or not, there might be another notion, also representable in INFL, whose status is clearer; and analyzing that notion would give us the answer regarding time.

2.2. Location

Time is often considered together with space, as two sides of the spatiotemporal coin. Recanati (2007a) argues that space, or location, is, just like time, also not an argument of the verb. He writes:

> There is no temporal or locative argument in the lexical entry for a verb like 'dance' (in contrast to verbs like 'arrive' or 'last', which take a locative and temporal argument respectively) ... Furthermore, times are introduced into the

logical form at the sentential level via the tense of the verb: this distinguishes times from locations, since there is no counterpart to tense in the spatial domain. (135)

In Recanati's view, then, the difference between time and location is not conceptual, but linguistic: time is expressed by tense, location is not. So, in a sense, the difference between tense and location is just an accidental property of English (and similar languages). If this is the case, we would expect to find languages where what corresponds to tense expresses location rather than time; moreover, we would expect location in these languages to behave in the same way that time behaves in English. This prediction is, in fact, borne out.

Ritter and Wiltschko (2005, 2009, 2014) discuss Halkomelem, a Salish language from Southwestern British Columbia and Northern Washington. They argue that this language is tenseless, and that, rather than tense, INFL indicates location: it can be filled either by the proximate í ('here') or the distal ni? ('there'). For example:

(9) a. ní? qw'eyílex tú-tl'ò
DIST dance he
'He is/was dancing (there).'
b. í qw'eyílex tú-tl'ò
PROX dance he
'He is/was dancing (here).'

Since, in Halkomelem, location is in INFL, which is outside the VP, it follows that location is not a syntactic argument of the verb. But what about the semantics of location? Does location behave like tense semantically?

A natural way to consider the question is to look at the semantic phenomenon of scope ambiguity. It is well known that tense can lead to scope ambiguities. For example:

(10) (during World War II) The British Prime Minister was a man.

Under one interpretation, tense takes narrow scope, and the sentence means that the *current* British Prime Minister (Theresa May) was a man in the past; under this reading, (10) is of course false. But when tense receives wide scope, (10) means that there was a time in the past when Britain had a male Prime Minister. Now the sentence is clearly true, since there were numerous times, e.g. during World War II, when the British Prime Minister was a man.

The two readings can be represented by the following logical forms:[6]

(10′)   a. ∃x (**pm**(x) ∧∀y(**pm**(y)→y=x) ∧ P̲ **man**(x))   (false)
        b. P̲ ∃x (**pm**(x) ∧∀y(**pm**(y)→y=x) ∧ **man**(x))   (true)

Now we should ask ourselves the following question: Do we get the same type of ambiguity with location in Halkomelem?

There is some evidence that the answer is yes. In Halkomelem both the distal and proximate auxiliaries come in two variants. The meanings of the auxiliaries and their variants are quite similar; for example, both (11a) (with the distal *niʔ*) and (11b) (with its variant *naʔət*) have the same English translation: 'The whale has been speared'.

(11)   a. niʔ   xʷə-s-θəθeq'   tθə   qʷənəs
              there   be-speared   the   whale
          b. naʔət   xʷə-s-θəθeq'   tθə   qʷənəs
              there   be-speared   the   whale

There are, however, subtle differences between the two sentences. What is relevant to our case here is that *niʔ* prefers the narrow scope reading of the distal operator, while *naʔət* prefers the wide scope reading. Citing a consultant, Gerdts (2010) points out that (11a)

> implies that the whale is in view but tells nothing about the event of spearing. The spearing could have happened elsewhere and the whale floated up to where we see it now. In contrast, [(11b)] is used when the speaker is pointing out to the [addressee] the actual location where the spearing took place and it is just over there.

Just as with the ambiguity of English tense, these two interpretations correspond to different scopal relations: while *naʔət* requires the existence of an individual that satisfies the predicates *be speared* and *be a whale* at the same location, the *niʔ* requires the existence of an individual that satisfies these predicates at two locations.[7]

For simplicity, and without judging the issue, I will assume a distal operator D̲, such that D̲ϕ indicates that ϕ is satisfied far away.[8] We can now represent the ambiguity as follows:

---

[6] For simplicity, and without prejudging the issues, I choose a Russellian interpretation of the definite determiner, and an operator theory of tense. Of course, this scope ambiguity can also be represented by other treatments of definites and, which is more pertinent to our concern here, using a quantificational theory of tense, which, in fact, later in the paper I will argue for.

[7] Unlike English tense, however, the two locations are allowed to be equal. The reason is that the while unspecified tense in English defaults to the present, unspecified location in Halkomelem defaults to the distal – see Ritter and Wiltschko (2009).

[8] We will be more precise concerning the meanings of the distal and proximate in Section 3.2 below.

(12) a. ∃x (**speared**(x) ∧ <u>D</u> (**whale**(x)∧
 ∀y (**whale**(y)→y=x)))
'The whale over there was speared'
b. <u>D</u> ∃x (**speared**(x) ∧ **whale**(x)∧
 ∀y (**whale**(y)→y=x))
'The whale was speared over there'

This ambiguity mirrors the scope ambiguity of tense.[9] Note that while the same tensed sentence in a language like English has two different interpretations, in Halkomelem the two interpretations are indicated by different auxiliaries, each preferring a different scope. But the phenomenon of different lexical items preferring different scopes is not unheard of. For example, the English modals *must* and *have to* arguably mean the same, except that the former takes wide scope with respect to negation, whereas the latter takes narrow scope: *Mary mustn't leave* means that Mary is under an obligation to not leave, while *Mary doesn't have to leave* means that Mary is not obligated to leave (Yanovich 2013).

We have seen that both tense and location behave syntactically the same: they occupy the same syntactic position, and are not arguments of the verb. We have now seen that they also behave semantically in the same way, leading to the same scope ambiguities. It therefore follows that either both are semantic arguments, or neither are. But which is it?

It may be easier to answer the question if we consider yet another notion: possible worlds. If possible worlds can be shown to behave linguistically like time and location, this would be evidence that either all three are semantic arguments of the verb, or all three are not. And, indeed, a language where INFL indicates modality is attested.

2.3. Possible world

Kĩsêdjê is an Amazonian language in which INFL encodes modality rather than tense (Nonato 2014). Nonato argues that, in this language, INFL indicates a relation between the world of evaluation *w*, and the world of the utterance w_0.

---

[9]Gerdts's own explanation of the facts does not use the notion of scope ambiguity, but is compatible with it.

Nonato identifies six particles that can occur at INFL, and describes their meanings in modal terms. First, let us consider what he calls the non-future particles:

1. *hẽn* (factual non-future)
   hẽn ϕ indicates simply that ϕ is true in the actual world. Thus, it indicates that the world of evaluation is the actual world: $w = w_0$
2. *waj* (inferential non-future)
   waj plays a role similar to epistemic modals in English. The world of evaluation is a world compatible with what is known in w_0
3. *man* (witnessed)
   man ϕ indicates that the speaker directly witnesses the truth of ϕ: w = a world compatible with what is witnessed in w_0
4. *arân* (counterfactual)
   In the counterfactual 'if ψ were true then *arân* ϕ', the particle indicates that w = the closest world to w_0 where ψ is true (cf. Stalnaker 1968)

Nonato also discusses two particles that he calls future particles. Does this negate his claim that Kīsêdjê is tenseless? Not necessarily, since future can be considered a modal rather than a tense (Abusch 1985; Enç 1987). We could, perhaps, conceive of the future as a modal indicating *predictions* (cf. Giannakidou and Mari 2018). Using such a definition, it is easy to characterize the two future particles as modals:

5. *kê* (factual future)
   kê ϕ expresses the prediction that ϕ will occur in the future: w = a world compatible with the predictions in w_0
6. *kôt* (inferential future)
   kôt is a combination of an epistemic modal and future, hence w = a world compatible with what is known and the predictions in w_0.

We have seen three notions that can be linguistically represented in INFL: time, location and possible worlds. Hence, we are forced to the conclusion that either all three are arguments of the verb, or all three are not. Perhaps one could still argue for the first position; but there is a fourth notion that is extremely hard to consider an argument of the verb: person.

2.4. Person

Consider a sentence such as

(13)   I lend you books.

The indisputable arguments of the verb are, of course, *I*, *you* and *books*. These arguments are in the first, second and third person, respectively. In many languages, the person of some of the arguments is reflected morphologically by the verb. For example, in English, if the subject were third rather than first person, the verb would be expressed as *lends* rather than *lend*. But the person of the argument, distinct from the argument itself, is not considered an additional argument of the verb. In a sentence such as (14), the verb does not have two arguments, one for the person snoring and another indicating the fact that this person is neither the speaker nor the addressee.

(14)   She snores

Interestingly, there are languages where INFL indicates person. Déchaine and Wiltschko (2014), Ritter and Wiltschko (2014), and Wiltschko (2014) discuss Blackfoot, an Algonquian language spoken in Southern Alberta and Northwestern Montana. In this language, verbs are not inflected for tense, but INFL contains information about person. Specifically, 'The order suffix –*hp* is used in root indicative clauses to signal that at least one participant of the reported event is also an utterance participant, i.e. a local (1st or 2nd) person' (Ritter and Wiltschko 2014, 9). The suffix -*hp* contrasts with -*m*, which indicates that all participants are non-local, i.e. third person.[10]

For examples, consider the sentences in (15), from Bliss (2013).[11]

(15)   a.   nit-ssim-atoo-**hp**-wa
            1-smell-TI-**hp**-PROX
            'I smelled it.'
       b.   kit   ssim   atoo   **hp**   wa
            2-smell-TI-**hp**-PROX
            'You smelled it.'
       c.   ii   ssim   atoo   **m**   wa
            IC-smell-TI-**m**-PROX
            'S/he smelled it.'

Note that the verb has two arguments: the smeller and the thing smelled; the suffixes -*hp* or -*m* are not arguments themselves, but are used to

---

[10] The suffix -*m* is often left null. The interpretation of null INFL is a fascinating topic, but it lies beyond the scope of this paper.
[11] 1 = first person, 2 = second person, TI = transitive inanimate, INAN = inanimate, PROX = proximate, IC = initial change.

indicate whether or not one of the actual arguments (the smeller and the thing smelled) is in the first or second person.

The important point for us is that -*hp* and -*m* fill the INFL position; but there is no conceivable way to construe them as arguments of the verb: they do not add any information about the event, let alone additional participants. Let me be clear about what I am saying here. I am not saying that participants in the event cannot be arguments of the verb: on the contrary, the smeller and the object that is smelled clearly *are* such arguments. But what I *am* saying is that *person*, as a category, is not an argument. Events have a number of characteristics: time, location, world, etc. It is under debate whether these characteristics are arguments of the verb or not; indeed, this is precisely the issue discussed here. But what is *not* under debate is whether the *person* of the participants – whether they are the speaker, addressee, or somebody else – is itself a characteristic of the event. Clearly, it is not, hence it could not conceivably be an argument of the verb.

So, we have found a case where, clearly, the content of INFL is not an argument of the verb. Since there is no reason whatsoever to assume that INFL is somehow sensitive to the content it carries and behaves differently with different contents, it follows that INFL, in general, is not an argument of the verb. From which it follows that all the notions representable in INFL are not arguments: time, location and possible world are not arguments of the verb.

It needs to be reiterated, however, that this does *not* mean that the operator approach has been vindicated; quite the contrary, I am going to use these and other conclusions to argue for the quantificational view instead.

3. Formalizing the quantificational approach

3.1. Time

If time is not an argument of the verb, this would seem to provide an argument for operators as opposed to quantifiers. Consider a simple sentence such as (16).

(16)   John sang

Under the operator approach, its representation would be something like (17).

(17)   P **sing(j)**

This logical form contains the past operator P, but, crucially, no temporal variable as an argument of the verb.

In contrast, the quantificational approach predicts something like the following as the logical form of (16) (where t_0 is the time of evaluation, usually the present).

(18) $\quad \exists t < t_0$ **sing'**(j,t)

Here, time is clearly an argument of the verb, apparently contradicting our finding from the previous section.

Things are not so simple, however. Recanati (2002) argues that the quantificational approaches can be made compatible with a theory of tense etc. as non-arguments.

He proposes that tense introduces what he calls *variadic functions*. These are functions that shift the type of the verb, so as to raise its adicity and allow an additional argument, and introduce a restricted quantifier binding this variable. The nature of the argument is determined by the linguistic construction. For example, tense introduces a temporal variable: 'temporal information is contributed by the tenses, so there is a temporal argument in the resulting logical form' (Recanati 2007a, p. 135); modal operators introduce world variables: '"Necessarily it rains" can be represented… in the same way, by applying to the modally neutral sentence "It rains" a sentence operator which explicitly quantifies over possible worlds and can be rendered as: "for every world *w*, in *w* it is the case that"… the variable "*w*" is introduced by the variadic operator' (Recanati 2007b, p. 87); a locative expression contributes a location variable: 'the prepositional phrase "in Paris" contributes not only the variadic function, but also the argument (Paris) which fills the extra argument-role' (Recanati 2002, pp. 319–320); and so on.

In the case of (16), its initial logical form is tenseless: it contains no temporal argument, but no other indication that the event occurred in the past:

(19) $\quad$ **sing**(j)

Only after the introduction of the variadic function does the verb **sing** shift to **sing'**, so as to have a temporal argument, bound by a quantifier, as in (18).

Therefore, according to both types of logical form, the operator and the quantificational, the initial logical form contains no temporal argument.

How, then, can we decide between the two theories? And, in particular, can linguistics help here as well?

Before answering this question, let us again consider what logical form the variadic functions would introduce in the case of the other notions considered here: location, world and person.

3.2. Location

The starting point for formalizing the distal and proximate will be Zwarts's (1997) vector-based semantics for *near*. For Zwarts, (20) means that the location of Jan is at the endpoint of a vector originating at the church with a length less than a contextually determined standard.

(20)   Jan is near the church

This interpretation is represented formally in (21), where r is the standard, **space**(x) is the set of vectors originating at the location of x, and **loc**(x,v) indicates that x is located at the endpoint of vector v.

(21) $\exists v \, (v \in$ **space**$([[\text{the church}]]) \wedge |v| < r \wedge$ **loc**$(j,v))$

What about the proximate, as in (9b), repeated below?

(9b)   í   qw'eyílex tú-tl'ò
        PROX   dance   he
        'He is/was dancing (here).'

After the introduction of a variadic function, the predicate **dance** will be shifted to **dance'**, which has an additional argument, indicating location. And (9b) will be true if there is at least one location where his dancing takes place, and which is the end point of a vector starting with the location of the speaker whose length is less than some standard. Formally:

(9b') $\exists l \, \exists v \, (v \in$ **space**$(l_0) \wedge |v| < r \wedge$ **loc**$(l,v) \wedge$ **dance'**$(\text{he},l))$

The distal will be formalized along similar lines.

3.3. Possible world

Take, for example, the following Kĩsêdjê, sentence, which uses the inferential non-future particle, *waj*:

(22)   Waj ngô-thyk   ta ta
        waj coffee-NOMINATIVE   stand
        'There must be coffee (left).'

Let us take **left(c)** to indicate the proposition that coffee is left. After type shifting it becomes **left'(c,w)**, indicating that coffee is left in world w. Let $K(w)$ be the set of worlds compatible with what is known in w, and let us assume the standard (and perhaps simplistic) analysis of epistemic necessity as universal quantification over $K(w)$. Then, after the application of the appropriate variadic function, the logical form of (22), under the quantificational approach, would be:

(23) $\forall w \in K(w_0)$ **left'(c,w)**

3.4. Person

The last concept to be discussed is person. Let us reconsider (15), repeated below:

(15) a. nit-ssim-atoo-**hp**-wa
      1-smell-TI-**hp**-PROX
      'I smelled it.'
   b. kit ssim atoo **hp** wa
      2-smell-TI-**hp**-PROX
      'You smelled it.'
   c. ii ssim atoo **m** wa
      IC-smell-TI-**m**-PROX

'S/he smelled it'. Let **smell(x,y)** indicate the proposition that x smells y. Type shifting then introduces a person variable, resulting in **smell'(x,y,p)**. How do we interpret this new predicate? Since, as said above, by no stretch of the imagination is person an argument of the verb, the answer is not immediately clear.

The standard account of the semantics of person (e.g. Dowty and Jacobson 1989; Schlenker 2002; Sauerland 2003) takes it to be presuppositional: a first-person pronoun presupposes that its referent is the speaker, a second person pronoun presupposes that it is the addressee, and a third person presupposes that it is neither. I am going to assume Heim's (2008) account, which uses Karttunen and Peters's (1979) two-dimensional theory of presupposition. The meaning of each expression has two dimensions: one is its standard truth conditional meaning (which, using more recent terminology, is usually called the *at-issue* meaning), and the second is its presupposition, or, as Karttunen and Peters call it, its Conventional Implicature (CI).

I suggest that type-shifting **smell**(x,y) into **smell'**(x,y,p) affects only the CI level, to presuppose that x or y are equal to p. In contrast, the at-issue level remains unaffected. Specifically, the interpretation of **smell'**(x,y,p) is:

(24) At-issue: **smell**(x,y)
   CI: x=p ∨ y=p

The -**hp** particle indicates that at least one of the arguments of the verb is the speaker or addressee. If s_0 is the speaker and a_0 is the addressee, the logical form of (15a) will be:

(15') a. ∃p∈{s_0, a_0} **smell'**(s_0,y,p)

Specifying this in terms of the at-issue and CI levels:

(15") a. At-issue: ∃p∈{s_0, a_0} **smell**(s_0,y)
   CI: ∃p∈{s_0, a_0} (x=p ∨ y=p)

The quantification at the at-issue level is vacuous, so it simply reduces to the statement that the speaker smelled it: **smell**(s_0,y). At the CI level we get the presupposition that at least one of the arguments is the speaker or addressee; this is the desired interpretation.

The interpretation of (15b) is similar. Its logical form is:

(15') b. ∃p∈{s_0, a_0} **smell'**(a_0,y,p)

Cashing it out we get the same CI level:

(15") b. At-issue: ∃p∈{s_0, a_0} **smell**(a_0,y)
   CI: ∃p∈{s_0, a_0} (x=p ∨ y=p)

This is the desired interpretation.

Things are slightly more complicated with (15c). The -**m** particle is, in a sense the negation of -**hp**: it says that the two arguments are *not* the speaker or addressee. The problem is that the CI level is usually assumed to be unaffected by negation: to use the common terminology, negation is a hole.

However, Karttunen and Peters note that sometimes the CI *can* be negated. They consider the following example:

(25) Even Bill likes Mary.

This sentence presupposes that 'Of the people under consideration, Bill is the least likely to like Mary'. While this presupposition cannot be negated by using standard negation, it *can* be negated, for example by saying 'Well yes, he does like her; but that is just as one should expect' (12).[12] Potts (2005) notes:

> Most natural language expressions for objecting to utterances target at-issue types ... But there are ways to get at the CI dimension. Karttunen and Peters (1979) observe that 'Well, yes, but ... ' is likely to indicate that the CI content is going to be disputed (12). Other strategies include 'Wait. I agree, but ... ' and even 'True, but ... '. The existence of these alternative strategies is a vindication of the multidimensional approach. It is impossible to make sense of a reply of the form 'True, but ... ' in a system in which sentence meanings have just one semantic dimension. (50–51)

Let $\sim$ indicate the type of negation that targets the CI level.[13] Then, we can represent the logical form of (15c) as follows:

(15')   c. $\forall p \in \{s_0, a_0\}$ $\sim$**smell**'(x,y,p)

Decomposing this into the at-issue and CI dimensions, we get:

(15")   c. at-issue: $\forall p \in \{s_0, a_0\}$ $\sim$**smell**'(x,y)
           CI: $\forall p \in \{s_0, a_0\}$ $\sim$(x=p $\vee$ y=p)

The quantification in the at-issue dimension is vacuous, and the $\sim$ operator does not apply to this level, so the at-issue level is simply equivalent to **smell**(x,y). But the CI level is different: the quantification is meaningful, and $\sim$ does apply at this level, so we get the desired interpretation: none of the arguments is equal to the speaker or addressee.

4. Indexicality and the quantificational approach

4.1. INFL as indexical

We have seen that languages vary with respect to which notion is expressed by INFL: in most languages it is tense, but in other languages it is modality, location, or person. But why does INFL express these notions? What is its role?

Hale (1986) discusses a two-way opposition between what he calls *central coincidence* and *non-central coindcidence*. Coincidence can be defined as a relation between figure and the ground with which it is compared. Central coincidence holds when the figure is identical to the ground or is included in it. Hale's main point is that this conceptual relation is

---

[12]See Umbach (2001) for more on the negation properties of *Yes, but*.
[13]This is not to be confused with Horn's (1989) *metalinguistic negation*. At any rate, it is doubtful that metalinguistic negation can deny presuppositions (Cohen 2006).

represented in the grammar of natural languages, and, moreover, is represented universally. Interestingly, according to Hale, 'the theme can be articulated in informal prose as follows: it is the definition of spatial, temporal, and identity relations in terms of "central" versus "non-central" (or "terminal") coincidence' (238). Note that all relations mentioned by Hale – spatial, temporal and identity (person) – are attested in INFL in some language. He could easily have added the fourth category – modality.

Ritter and Wiltschko (2009, 2014) propose that INFL expresses this universal opposition and uses it to *anchor* the event situation to the utterance situation. In languages like English where INFL instantiates tense, the anchoring is temporal: tense tells us *when* the reported event happened – *now* or *not now*. In a language like Halkomelem, it tells us *where* it happened: *here* or *not here*; in a language like Blackfoot *who* the participants in it were – *us* or *not us*. In other words, this proposal treats INFL as a sort of indexical. Indeed, as far as is known, in no language does INFL instantiate a non-indexical notion such as number.[14]

4.2. No INFL Without Indexical

Note that all the logical form from the previous section involve indexical elements: t_0 is an indexical referring to the present time, w_0 is an indexical referring to the actual world, l_0 is an indexical referring to the location of the utterance, and s_0 and a_0 correspond to the person indexicals *I* and *you*, respectively. This fact will become crucial in deciding between the operator and quantificational view.

Consider the linguistic predictions of the quantificational representation of tense. After the application of the variadic function, the logical form crucially involves t_0, corresponding to the indexical *now*. What would happen in a language that simply lacks an indexical that refers to times, a language that doesn't have a *now*?

In such a language, the application of the variadic function would be uninterpretable. Therefore, the quantificational approach predicts that such a language would have to be tenseless: INFL will have to represent some other feature besides time.

In general, different languages use different types of indexical. We therefore make the following prediction: if a language lacks an indexical

---

[14]In a language like Kĩsêdjê, INFL tells us in which *world* the reported event happened – the actual word or not. As we will see in Section 4.6 below, this, too, is an indexical notion.

for a certain notion, INFL cannot indicate this notion. Since this principle concerns the connection between indexicality and INFL, let us call it the *No INFL Without Indexicality* principle, or NIWI.

Note that we predict an implication in one direction only: if a language does not have, say, a temporal indexical, then INFL cannot indicate time, and the language will have to be tenseless. But if a language does have such an indexical, it still does not follow that INFL will indicate time: it might indicate world, location or person, provided the language has indexicals for these notions.

What about the operator view? NIWI would, as far as I can see, be compatible with it, but crucially, it does not follow from it, since indexicals play no important role in the operator approach. That is to say, if it turns out that NIWI holds, it would, under the operator view, be just a coincidence, an accidental generalization that has no explanation.

The NIWI principle can be empirically tested: let us consider a number of languages, and see whether we can find evidence supporting or refuting NIWI.

4.3. Blackfoot

Ritter and Wiltschko (2005) point out that there are no temporal indexicals in Blackfoot. They grant that Blackfoot has words that *appear* to be temporal indexicals; for example, *matúnnii*, which had been glossed as the indexical *yesterday* by Frantz and Russell (1995). However, Ritter and Wiltschko demonstrate that it is better interpreted as the non-indexical *the day before*. For example, (26) can be interpreted as *Mary saw him yesterday*; but if preceded by a sentence such as *I saw John one day last week and* ..., can only mean that Mary saw him the day before I saw him, i.e. she saw John last week, rather than yesterday.

(26) Namyááni náínooyiwai matúnnii
  this-Mary see-him day.before

Since Blackfoot does not have temporal indexicals, NIWI predicts that this language will be tenseless, i.e. INFL will not indicate time.

As we have seen above, Blackfoot does have person indexicals, such as *nit* 'I' and *kit* 'you'. Therefore, NIWI also predicts that it is possible in that language for INFL to indicate person – and this is, indeed, the case.

We must use caution, though. Kaplan (1989) conceived of indexicals as elements whose reference is determined only by the context of the utterance situation, and does not change when they are embedded in the scope of logical operators. However, Schlenker (2003) has demonstrated that this is not, in general, the case: there are languages where indexicals can be *shifted* by embedding them. For example, in English, (27) can only mean that John said that the speaker is a hero; but, in Amharic, the corresponding sentence can mean that John said that he – John – is a hero.

(27)   John said that I am a hero.

Many more examples of such languages have been discovered since Schlenker's paper. An important constraint that shifted indexicals seem to obey is the Shift-Together principle (Anand and Nevins 2004): in an embedded clause, either all indexicals shift, or none of them do.

Could it be argued that Blackfoot is another such language, and that it does have temporal indexicals, but that they are shifted indexicals? If this were the case, it would not contradict the NIWI principle, but neither would it confirm it.

4.4. Nez Perce

Unfortunately, the possibility of shifted indexicals in Blackfoot has not been studied, so we cannot answer this question. But the issue has been taken up with another interesting language – Nez Perce, a Sahaptian language spoken in northwestern United States.

Consider the following example, from Deal (2014):

(28)   Beth-nim   hi-hi-n-e   'e-wewkuny-e   sepehitemenew'etuu-ne
       Beth   said-to-me   I-met   teacher
   Literally: 'Beth told me I met the teacher'

Sentence (28) means that Beth told me that *she* met the teacher. Deal notes that (28) gets this interpretation even if Beth does not know that the person she met was a teacher, precluding the possibility that this is direct quotation, and making it clear that the indexical is indeed embedded. Deal also shows that Nez Perce obeys Anand and Nevins's (2004) Shift-Together principle.

What is pertinent to our discussion here is Deal's argument that Nez Perce, like Blackfoot, does not have temporal indexicals. Nez Perce does have expressions that are often translated as temporal indexicals, e.g. *kii taaqc* 'today' and *watiisx* 'tomorrow', but these are not really indexical.

They can shift in an embedded context, and, moreover, unlike shifted indexicals, they do not have to obey the Shift-Together constraint:

(29) naaqc k'ay'x-pa weet 'aayat hi-i-cee-ne ki-yu' kii taaqc 'itq'o watiisx

one week-ago QUESTION lady said I-go same-day or one-day-later
Literally: 'One week ago, did the lady say that I would go the same day or one day later?'

The crucial point is that it is possible for only one of the temporal expressions to shift, while the other one does not. Suppose (29) is uttered on the 20th day of the month. Then it can be interpreted as a question about what the lady said a week ago, i.e. on the 13th: did she say that she would go the same day as the *reported* speech event (i.e. the 13th), or one day later than the day of *utterance* (i.e. on the 21st)? This fact shows clearly that these expressions are not indexical, not even shifted ones.

Deal argues that, in contrast, Nez Perce does have locative indexicals. These indexicals can shift, but they obey the constraint Shift-Together. Consider the following sentence, which is uttered in a town called Lapwai (Lewiston is the closest major city):

(30) miniku cewcewin'es hi-i-caa-qa Simiinikem-pe hi-muu-no'qa ki-nix
which phone they-said in-Lewiston can-call from-here
met'u weet'u hi-muu-no'qa ko-n'ıx?
but not can-call from-there
Literally: 'Which phone did they say in Lewiston that it can call from here but not from there?'

Sentence (30) can be interpreted as a question about a phone that can call from Lapwai (the 'here' of the utterance) but not from Lewiston (the 'there' of the utterance). Alternatively, (30) can be about a phone that can call from Lewiston (the 'here' of the embedded context) but not from Lapwai (the 'there' of the embedded context). That is to say, either both locative indexicals shift, or none do. Crucially, it is impossible for only one of the indexicals to shift while the other does not; (30) cannot get a (contradictory) reading where the phone can and cannot call from Lapwai, or can and cannot call from Lewiston.

Since Nez Perce does not have temporal indexicals, but does have locative indexicals, the NIWI principle predicts that Nez Perce is tenseless, and that INFL could indicate location.

Interestingly, Deal (2008) claims that Nez Perce actually does have tense: specifically, there are suffixes that supposedly indicate past or future. However, it is not clear that these suffixes really indicate tense, rather than aspect.

Linguists usually distinguish between tense, which is 'the grammaticalized expression of location in time' and aspect, which is about the 'internal temporal constituency' of a situation (Comrie 1985, 9–10). While this theoretical distinction is reasonably well understood, in practice, the identification of a particular construction as indicating tense or aspect is notoriously difficult (Comrie 1976, 1985; Tedeschi and Zaenen 1981; Hopper 1982; Dahl 1985; Palmer 1986; Bhat 1999; de Saussure, Moeschler, and Puskás 2007, Wiklund 2007).

Crucially, the suffixes Deal discusses only occur with some aspectual categories, but not others. This is rather a strange behavior for a tense particle, because every situation, regardless of its internal temporal constituency, has a time. Therefore, it is plausible that these suffixes actually indicate aspect rather than tense. The question is not easy to determine conclusively. Indeed, Velupillai (2016) admits that the issue is 'difficult to decide' (109),[15] and even Deal herself notes for one of the suffixes that it 'behaves like an aspect' (2008, 231). I therefore conclude that Deal has not made a convincing case that these suffixes really indicate tense, and that Nez Perce is, indeed, tenseless, as predicted by NIWI.

What about location? It turns out that Nez Perce, like Halkomelem, contains a proximate (-*m*) and a distal (-*nki*) particle. These particles indicate distance from the speaker. Consider the following minimal pair, from Deal (2010):

(31) a. hi-weqi-se-m
it-rain-ing-here
'It is raining here'
b. hi-weqi-see-nki
it-rain-ing-there
'It is raining over there'

What is the syntactic position of these particles? Because Deal believes that, in Nez Perce, INFL indicates tense, she places the location particles

---

[15]Although, eventually, she has 'chosen to assign the value "three tenses" to Nez Perce'.

immediately above INFL. However, we have seen reasons to believe that, in fact, Nez Perce, is tenseless; if this is, indeed, the case, the locative particles ought to be in INFL. This is exactly as predicted by the NIWI principle.

4.5. Halkomelem

We have seen that, in Halkomelem, INFL encodes location. According to NIWI, this language ought to have locative indexicals, which indeed it does, for example, *ikwelo* 'here', as the following examples, from Russell, Phillips, and Williams (2016) demonstrate:

(32) a. itet ikwelo
           he-slept here
'He slept here'
      b. me xwe ikwelo
           they came here
'They came here'.

4.6. Kĩsêdjê

Kĩsêdjê has a modal indexical – the dubitative *jantã*, exemplified by the following text, from de Sousa (2009):

(33) a. adjikatorotá na wi mbaj kêrê wâtân adjinhihwêrê
         'Of our origin, we do not even know what made us people:'
      b. kajkwa-kãm-wapãm na adjinhihwêt jantã
         'our father in heaven who made us, maybe'

The English translation of *jantã* as 'maybe' is somewhat misleading: *jantã* is actually a determiner, and, in Kĩsêdjê, the determiner follows the noun phrase, so the entire (33b) is a referential expression, referring to our father in heaven who made us. The dubitative indicates that the existence of this referent in the actual world is in doubt, so (33b) must refer to an individual in some other world. Note that the doubt is expressed by the speaker; if the others did not know who made them people, but the speaker did know, the use of the dubitative would not be acceptable. Hence, *jantã* is a modal indexical.

As predicted by the NIWI, since Kĩsêdjê has modal indexicals, it may have a modal INFL, which, as we have seen, it does.

4.7. English

Lewis (1970) and Kaplan (1989) claim that *actual* or *actually* are modal indexicals in English. However, they seem to have in mind a technical sense of the word as used by philosophers, and not the regular use of the English word.

One way to see this is to note, as Chalmers (2011) does, that for Lewis and Kaplan (33) holds.

(34)   *Actually Φ* entails *necessarily actually Φ*

Chalmers points out that 'one can deny that the English word "actually" satisfies' (34), although he concedes that 'it is hard to deny that there is a technical term that works this way' (411).

Davis (2015) notes: 'The indexicality thesis has not spread to linguistics, though. The entries on indexicals in the two standard encyclopedias of linguistics do not give "actual" or "actually" as examples' (470). Indeed, linguistic studies of *actually* (e.g. Oh 2000; Smith and Jucker 2000; Taglicht 2001; Cheng and Warren 2002) treat it in terms of contrast or discrepancy.

So, we can conclude that English does not have modal indexicals; the NIWI principle predicts that it cannot have a modal INFL, which indeed it does not.

4.8. Null indexicals?

Therefore, the behavior of all five languages we have considered turns out to be compatible with NIWI. And this, in turn, provides support for the quantifiers approach, at the expense of operator-based theories, for which this generalization would be an inexplicable mystery.

But could it be argued that, in a language where an indexical of a certain sort is not attested, such an indexical exists after all, but it is null?

The answer is no. And the reason is that even if null indexicals exist, they behave differently from overt indexicals, in that they are not subject to the Shift-Together constraint. Consider the following example from the language Mishar Tatar (Tyler 2015):

(35)   min Marat-ka   [*pro* **sine**   sü-m-i-**seŋ**   diep] at'-tɤ-m
       I   Marat-DAT [**you-ACC**   love-NEG-ST-**2SG** C]   tell-PST-1SG
       'I told Marati that hei doesn't love you.'
       (lit. 'I told Marati that youi don't love you')

While the overt *sine* 'you' is not shifted, the null *pro* is.[16] Tyler notes that this phenomenon is observed in other languages (e.g. Şener and Şener 2011), applies it to English, and turns it into a general, cross-linguistic principle. We can conclude, then, that an appeal to null indexicals would not help.

5. Conclusions

We started our investigation from a philosophical question: are times, worlds, etc. arguments of the verb? We turned to linguistics for an answer. We have seen that, in different languages, all these elements can fill the same syntactic position – INFL – and that they behave semantically the same, in the sense that they give rise to the same scope ambiguities. We therefore concluded that either all of them are arguments, or neither is. And while one might conceivably entertain the notion that time, world and location are arguments, it is very hard to argue that person is an argument. And if person is not an argument, it follows that none of the others are either.

Although this conclusion seemed, on the face of it, to support the operator view, we have seen that it does not: with the help of variadic functions, the verb can be type-shifted so that what is not an argument initially becomes an argument, thus accommodating the quantificational view.

But there is a crucial difference between the operator and the quantificational approaches: only the latter essentially requires indexicals of the type of variables that is quantified over. We therefore predicted the NIWI principle: No INFL Without Indexical. That is to say, in a given language, INFL cannot indicate a certain notion unless the language had an indexical of the sort of that notion.

NIWI has been confirmed by all the languages we looked at. Blackfoot and Nez Perce lack temporal indexicals, and consequently do not have temporal INFL. But Blackfoot does have person indexicals, and INFL, in this language, indicates person; and Nez Perce has locative indexicals, and so INFL is allowed to indicate location, and the same goes for Halkomelem. Kĩsêdjê has a modal indexical, and consequently can (and does) have a modal INFL; but English, which lacks a modal indexical, cannot.

The NIWI principle follows naturally from the quantificational approach. In contrast, this generalization would remain a mystery under the operator account. Under that view, why should there be any connection between

---

[16]See Podobryaev (2014) for more on this phenomenon.

indexicalty and the role of INFL? Hence, the quantificational approach provides a better, principled explanation of the data, and is therefore to be preferred.

If the arguments proposed in this paper are on the right track, these results have wider implications concerning the use of linguistic evidence in philosophical debates. We have seen that Recanati argues, on philosophical grounds, that elements like time and world are not arguments of the verb. We have seen that linguistic evidence, from a variety of languages, supports this claim.

Recanati also argues that the issue of whether time etc. are arguments, even if it is resolved, does not decide the wider question of whether the quantificational or operator view is correct, and that both views can be made compatible with the view that they are not arguments. But we have seen that linguistic evidence, does, in fact, provide evidence in favor of one of the views – the quantificational approach.

Acknowledgments

I would like to thank François Recanati for numerous thorough discussions and insightful suggestions. I also thank Betsy Ritter, from whom I learned a great deal about the syntax of functional categories.

Disclosure statement

No potential conflict of interest was reported by the author.

ORCID

Ariel Cohen http://orcid.org/0000-0002-0568-0792

References

Abusch, D. 1985. "On Verbs and Time." PhD thesis., University of Massachusetts, Amherst.
Anand, P., and A. Nevins. 2004. "Shifty Operators in Changing Contexts." In *Proceedings of SALT XIV*, edited by Robert B. Young, 20–37. Ithaca, NY: CLC.
Bhat, D. N. S. 1999. *The Prominence of Tense, Aspect and Mood (Studies in Language Companion Series)*. Amsterdam: John Benjamins.
Bliss, H. 2013. "The Blackfoot Configurationality Conspiracy: Parallels and Differences in Clausal and Nominal Structures." *PhD diss.*, UBC.
Chalmers, D. J. 2011. "Actuality and Knowability." *Analysis* 71 (3): 411–419.

Cheng, W., and M. Warren. 2002. "The Functions of Actually in a Corpus of Intercultural Conversations." *International Journal of Corpus Linguistics* 6 (2): 257–280.

Chomsky, N. 1986. *Barriers*. Cambridge, MA: MIT press.

Chung, S., and W. A. Ladusaw. 2004. *Restriction and Saturation*. Cambridge, MA: MIT Press.

Cohen, A. 2006. "How to Deny a Presupposition." In *Where Semantics Meets Pragmatics*, edited by K. Turner and K. von Heusinger, 95–110. Amsterdam: Elsevier.

Comrie, B. 1976. *Aspect: An Introduction to the Study of Verbal Aspect and Related Problems*. Cambridge: Cambridge University Press.

Comrie, B. 1985. *Tense*. Cambridge: Cambridge University Press.

Dahl, Ö. 1985. *Tense and Aspect Systems*. Oxford: Blackwell.

Davis, W. A. 2015. "The Semantics of Actuality Terms: Indexical vs. Descriptive Theories." *Noûs* 49 (3): 470–503.

Deal, A. R. 2008. "Events in Space." In *Proceedings of Semantics and Linguistic Theory (SALT)*. Vol. 18., edited by T. Friedman and S. Ito, 230–247. Ithaca: Cornell University.

Deal, A. R. 2010. "Topics in the Nez Perce Verb." University of Massachusetts Amherst dissertation.

Deal, A. R. 2014. "Nez Perce Embedded Indexicals." In *Proceedings of SULA 7: Semantics of Under-Represented Languages in the Americas*, edited by Hannah Greene, 23–40. Amherst: GLSA.

Déchaine, R. M., and M. Wiltschko. 2014. "Micro-variation in Agreement, Clause-Typing and Finiteness: Comparative Evidence from Blackfoot and Plains Cree." In *Papers of the 42nd Algonquian Conference*, edited by J R Valentine and M Macaulay, 69–101. Albany: SUNY Press.

de Saussure, L., J. Moeschler, and G. Puskás, eds. 2007. *Tense, Mood and Aspect: Theoretical and Descriptive Issues*. Vol. 17. Amsterdam: Rodopi.

de Sousa, M. S. C. 2009. "Três nomes para um sítio só: a vida dos lugares entre os Kĩsêdjê (Suyá)." IV congresso da Associação Portuguesa de Antropologia, Painel Convidado VII: Classificar: objectos, sujeitos, acções(in Portuguese).

Dowty, D., and P. Jacobson. 1989. "Agreement as a Semantic Phenomenon." In *Proceedings of the Fifth ESCOL*, edited by J. Powers and F. de Jong, 95–101. Ithaca, NY: Cornell Linguistics Club.

Enç, M. 1987. "Anchoring Conditions for Tense." *Linguistic Inquiry* 18: 633–657.

Şener, N. G., and S. Şener. 2011. "Null Subjects and Indexicality in Turkish and Uyghur." Proceedings of WAFL 7.

Frantz, D., and N. Russell. 1995. *Blackfoot Dictionary*. Toronto: University of Toronto Press.

Gerdts, D. 2010. "Agreement in Halkomelem Complex Auxiliaries." In *Proceedings of the International Conference on Salish and Neighbouring Languages (ICSNL) 44/45 (UBCWPL vol. 27)*, edited by J. Dunham and J. M. Lyon, 175–189. Vancouver: University of British Columbia.

Giannakidou, A., and A. Mari. 2018. "A Unified Analysis of the Future as Epistemic Modality." *Natural Language & Linguistic Theory* 36 (1): 85–129.

Glasbey, S. 1992. "Distinguishing Between Events and Times: Some Evidence from the Semantics of `Then'." *Natural Language Semantics* 1: 285–312.

Grimshaw, J. 1994. "Lexical Reconciliation." *Lingua* 92: 411–430.

Hale, K. 1986. "Notes on World View and Semantic Categories: Some Warlpiri Examples." In *Features and Projections*, edited by Pieter Muysken and Henk van Riemsdijk, 233–254. Dordrecht: Foris.

Heim, I. 2008. "Bound Pronouns and Morphology." Handout from a Talk at UBC, November.

Hopper, P. J., ed. 1982. *Tense-Aspect: Between Semantics and Pragmatics*. Amsterdam: Benjamins.

Horn, L. R. 1989. *A Natural History of Negation*. Chicago, IL: University of Chicago Press. Reissued 2001 by CSLI.

Jackendoff, R. 1992. *Semantic Structures*. Cambridge, MA: MIT press.

Kaplan, D. 1989. "Demonstratives: An Essay on the Semantics, Logic, Metaphysics, and Epistemology of Demonstratives and Other Indexicals." In *Themes from Kaplan*, edited by J. Almog, J. Perry, and H. Wettstein, 481–563. New York: Oxford University Press.

Karttunen, L., and S. Peters. 1979. "Conventional Implicature." In *Syntax and Semantics 11: Presupposition*, edited by C.-K. Oh and D. Dinneen, 1–56. New York: Academic Press.

Koopman, H., and D. Sportiche. 1991. "The Position of Subjects." *Lingua* 85: 211–258.

Lewis, D. 1970. "Anselm and Actuality." *Noûs* 4: 175–188.

Nonato, R. 2014. "Clause Chaining, Switch Reference and Coordination." *Doctoral diss.*, Massachusetts Institute of Technology.

Oh, S. Y. 2000. "Actually and in Fact in American English: A Data-Based Analysis." *English Language and Linguistics* 4 (02): 243–268.

Palmer, F. R. 1986. *Mood and Modality*. Cambridge: Cambridge University Press.

Podobryaev, A. 2014. "Persons, Imposters and Monsters." *Doctoral diss.*, MIT.

Potts, C. 2005. *The Logic of Conventional Implicatures*. Oxford: Oxford University Press.

Ramchand, G. 2004. "Time and the Event: The Semantics of Russian Prefixes." In *Nordlyd 32.2: Special Issue on Slavic Prefixes*, edited by P. Svenonius, 323–361. Tromsø: University of Tromsø.

Recanati, F. 2002. "Unarticulated Constituents." *Linguistics and Philosophy* 25 (3): 299–345.

Recanati, F. 2007a. "It is Raining (Somewhere)." *Linguistics and Philosophy* 30 (1): 123–146.

Recanati, F. 2007b. *Perspectival Thought: A Plea for (Moderate) Relativism*. Oxford: Oxford University Press.

Ritter, E., and M. Wiltschko. 2005. "Anchoring Events to Utterances Without Tense." In *Proceedings of WCCFL*. Vol. 24., edited by J. Alderete, C. Han, and A. Kochetov, 343–351. Somerville, MA: Cascadilla Proceedings Project.

Ritter, E., and M. Wiltschko. 2009. " Varieties of INFL: Tense, Location, and Person." In *Alternatives to Cartography*, edited by Jeroen van Cranenbroeck, 153–201. Berlin: Mouton de Gruyter.

Ritter, E., and M. Wiltschko. 2014. "The Composition of INFL." *Natural Language & Linguistic Theory* 32 (4): 1–56.

Russell, S., E. Phillips, and V. Williams. 2016. "Telling Stories in a Halq'emelem Conversation: Doing Beginnings and a Bit About Endings." In *Papers from the*

International Conference on Salish and Neighbouring Languages 51, University of British Columbia Working Papers in Linguistics 42.

Sauerland, U. 2003. "A New Semantics for Number." In *SALT 13*, edited by R. B. Youn and Y. Zhou, 258–275. Ithaca, NY: Cornell Linguistics Club.

Schlenker, P. 2002. "Indexicality, Logophoricity, and Plural Pronouns." In *Research in Afroasiatic Grammar II*, edited by J. Lecarme, 409–428. Amsterdam: Benjamins.

Schlenker, P. 2003. "A Plea for Monsters." *Linguistics and Philosophy* 26: 29–120.

Smith, S. W., and A. H. Jucker. 2000. "Actually and Other Markers of an Apparent Discrepancy Between Propositional Attitudes of Conversational Partners." In *Pragmatic Markers and Propositional Attitude*, edited by G. Andersen and T. Fretheim, 207–238. Amsterdam: John Benjamins.

Stalnaker, R. 1968. "A Theory of Conditionals." In *Studies in Logical Theory*, edited by N. Rescher, 98–112. Oxford: Blackwell.

Taglicht, J. 2001. "Actually, There's More to it than Meets the Eye." *English Language and Linguistics* 5 (1): 1–16.

Tedeschi, P., and A. Zaenen, eds. 1981. *Tense and Aspect (Syntax and Semantics 14)*. New York: Academic Press.

Tyler, M. 2015. "English Embedded Imperatives Have a Context-Shifting Operator." Handout of a Talk, CGG, 25, IKER.

Umbach, C. 2001. "Contrast and Contrastive Topic." In *Proceedings of the ESSLLI 2001, Workshop on Information Structure, Discourse Structure and Discourse Semantics*, edited by I. Kruijff-Korbayov and M. Steedman, 2–13. Helsinki: University of Helsinki.

Velupillai, V. 2016. "Partitioning the Timeline: A Cross-Linguistic Survey of Tense." *Studies in Language* 40 (1): 93–136.

Wiklund, A.-L. 2007. *The Syntax of Tenselessness: Tense/Mood/Aspect-Agreeing Infinitivals*. Berlin: Mouton de Gruyter. Studies in Generative Grammar 92.

Wiltschko, M. 2014. *The Universal Structure of Categories: Towards a Formal Typology*. Cambridge: Cambridge University Press.

Yanovich, I. 2013. "Four Pieces for Modality, Context and Usage." *Doctoral diss.*, Massachusetts Institute of Technology. Chapter 5.

Zwarts, J. 1997. "Vectors as Relative Positions: A Compositional Semantics of Modified PPs." *Journal of Semantics* 14 (1): 57–86.

Confessions of a schmentencite: towards an explicit semantics

Jonathan Schaffer

ABSTRACT
Natural language semantics is heir to two formalisms. There is the extensional machinery of explicit variables traditionally used to model reference to individuals, and the intensional machinery of implicit index parameters traditionally used to model reference to worlds and times. I propose instead a simple and unified extensional formalism – *explicit semantics* – on which all sentences include explicit individual, world and time variables. No implicit index parameters are needed.

> I concede this victory to the schmentencite: strictly speaking, we *do not need* to provide both context-dependence and index-dependence in the assignment of semantic values to genuine sentences. His victory is both cheap and pointless. I propose to ignore it. (Lewis 1980, 90)

Natural language semantics is heir to two formalisms. There is the extensional machinery of explicit variables evaluated through the assignment function and bindable by quantifiers, traditionally used to model reference to individuals; and there is the intensional machinery of implicit index parameters initialized from the speech context and shiftable by operators, traditionally used to model reference to worlds and times. So there are debates over the best machinery for times, and proposals to add more parameters for further matters such as locations and perspectives.

I think that the use of two formalisms was already a mistake. There is neither need nor motivation for kludging together both sorts of machinery, because the phenomena – individual, world and time reference in natural language – are relevantly similar, and the extensional and intensional formalisms – suitably developed – prove expressively equivalent.

Moreover, I argue that the extensional machinery is generally preferable. I thus defend what I call *explicit semantics*, on which all sentences include explicit world and time variables. These variables are articulated through syntactically obligatory mood and tense phrases, evaluated through the variable assignment, and bindable by modal and temporal quantifiers. The propositions expressed are simply true or false, without relativity to implicit points of reference (for world, time, or any other matter). I claim that the resulting style of semantics is not just simpler and more elegant, but also more constrained, and capable of delivering classical propositions: complete thoughts fit to bear truth-values.

Overall I aim to put explicit semantics on the map, as a positive and motivated view. Obviously, I cannot discuss every argument concerning world and time reference, nor consider every proposed parameter. My strategy is rather to show how the original arguments against extensional treatments of world and time fail, sketch an empirically motivated extensionalist alternative (fitting Lewis's 'schmentencite' view), and explore what may be gained from embracing explicit semantics as a whole. The skeptical reader should treat this as an invitation to explain why natural language semantics needs any intensional machinery whatsoever.

Overview: In §§1–4 I trace a historical arc through Montague and Kaplan, and on to eternal and then necessary contents, in which information needed for truth evaluation gets progressively shunted from the index into the proposition. Explicit semantics lies at the endpoint of this arc, when the index is emptied. In §§5–6 I compare explicit semantics as a whole to the mixed and purely implicit alternatives.

1. Montague on pragmatic languages, and the birth of index semantics

In the beginning, there was Montague, whose revolutionary idea is that natural languages can be modeled as formal systems via intensional logics. The pesky context-dependence of 'pragmatic languages' is treated by starting from an intensional language with an index for world and time parameters, and adding parameters for speakers, places, addressees and any other relevant contextual matters. This idea – shared by Montague (1968, 1970, 1973) and others such as Scott (1970), and the early Lewis (1970) – is perhaps most clearly expressed by Scott (1970, 151) as follows:

In general, we will have:

$$i = (w, t, p, a, \ldots)$$

where the index *i* has many coordinates: for example, *w* is a world, *t* is a time, $p = (x, y, z)$ is a (3-dimensional) position in the world, *a* is an agent, etc. All these coordinates can be varied, possibly independently, and thus affect the truth-values of statements which have indirect references to these coordinates.

So the starting point idea, from Montague, is that the semantic machinery is:

Thus consider the sentence:

1. I am writing

There are at least two aspects of context-dependence involved in 1. The first concerns the explicit indexical 'I'. The second concerns the world and time at issue, which – at least on the surface – does not seem explicit in the sentence. Index Semantics handles both aspects of context-dependence through the index.[1] Eliding over the compositional backstory, Index Semantics delivers the verdict that 1 is true at context *c* if and only if the index i_c is such that the agent of i_c is in the extension of the predicate 'am writing' at the world and time of i_c. This is a breakthrough: Montague's work opens up the vision of a precise, compositional treatment of natural language that can manage context-dependence.

There are three points worth highlighting. The first is that natural language semantics begins in the image of modal logic, and indeed with Montague's idea that the indices of intensional systems can manage all of the context-dependence of natural language. That proved to be a fruitful starting point. But it is an empirical question as to whether natural language really has the structure of an intensional system. (As I see it, the history of semantics I am tracing is the history of gradually erasing this fruitful mistake.)

Secondly, parameters were freely posited as needed. Thus Scott in the passage above proposes an open-ended list ('$i = (w, t, p, a, \ldots)$'), and

---

[1] As Lewis (1980, 85) would later tell the tale, the friend of Index Semantics (including his earlier self: Lewis 1970) had posited 'a happy coincidence' between these aspects of context-dependence. To which he adds: 'How nice'. Then: 'No; we shall see that matters are more complicated.'

Cresswell (1972, 8; cf. Lewis 1980, 87) notes that this way lies a coordinate for 'previous drinks', and that in general 'there is no way of specifying a finite list' of parameters for all contextually relevant information. (As I argue in §§5–6, explicit semantics adds useful constraints.)

Thirdly and most crucially, there were two guiding rationales for separating reference to individuals (via variables) from reference to worlds and times (via parameters). The first guiding rationale was that simple sentences like 1 were considered world and time neutral. Thus Kamp (1971, 231) says, with respect to times:

> I of course exclude the possibility of symbolizing the sentence by means of explicit quantification over moments... Such symbolizations... are a considerable departure from the actual form of the original sentences which they represent – which is unsatisfactory if we want to gain insight into the semantics of English.[2]

And likewise Recanati (2007, 66) says, with respect to worlds:

> Clearly, the possible worlds relative to which propositions are evaluated are not themselves represented in, or an aspect of, the propositions in question. Thus 'Brigitte Bardot is French' is true, with respect to a world w, iff Brigitte Bardot is French in w; but the sentence 'Brigitte Bardot is French' only talks about Brigitte Bardot and the property of being French. The world of evaluation is not a constituent of the content to be evaluated.

(As I argue in §§3–4, this perspective is natural but syntactically naive, ignoring the obligatory mood and tense features specifying the world and time at issue. For instance, in 'Brigitte Bardot is French', English plays the trick of packing the root verb $\sqrt{be}$, indicative mood and present tense all within the seemingly simple 'is'.)

The second guiding rationale was that natural languages were thought to be impoverished vis-a-vis world and time reference. It was seen that natural languages could track an indefinite collection of individuals, so this was modeled through an infinite stock of variables. But it was also thought (at first) that natural languages could only hold one world and one time perspective. Hence it was thought that world and time should be modeled by just a single world point and time point (Prior 1957). As Kratzer (2014, §5) says:

---

[2]Though Kaplan (1989a, 503) allows that 'An alternative [and more traditional] view is to say that the verb tense in S involves an implicit temporal indexical ... ' But he questions why we should adopt this view, when 'no modal indexical is taken to be implicit'. I think that Kaplan is right to keep world and time in parallel, but only wrong not to consider the prospect of a modal indexical more seriously (§4).

Quantification over worlds and times is treated differently from quantification over individuals, then. The distinction was made deliberately because it predicts differences that were thought correct at the time. Once an evaluation index is shifted, it is gone for good, and can no longer be used for the evaluation of other expressions. This constrains temporal and modal anaphora. Until the early seventies anaphoric reference to times and worlds in natural languages was believed to be constrained in precisely the way predicted by the evaluation index approach.

This is the story of the collapse of these two guiding rationales for separating reference to individuals from reference to worlds and times, and the emergence of symmetry between these referential systems. As Schlenker (2006, 510) says: '[T]he overwhelming evidence is that the semantic differences that were traditionally posited between the three ontological domains are largely imaginary.' Natural language semantics is heir to two formalisms, but these complications have outlived their rationales.

This is also the story of the (re-)emergence of *the proposition* lying between the sentence and its truth-value, and of pulling time reference and then world reference from the index into the proposition. Explicit semantics lies at the end of the story, when the index is finally emptied, and the proposition made complete.

2. Kaplan on indexicals, and Lewis on the schmentencite way out

Kaplan (1989a; though the material was circulating by 1971) argues that Montague Semantics conflates the evaluation of indexicals with the treatment of intensions, and thus separates the two aspects of context-dependence seen in 1. Kaplan does not question the background intensional formalism he inherits from Montague, and so retains an index with a world and time parameter <w, t>, and the concomitant treatment of modals and temporals as operators that shift implicit parameters.

But Kaplan's breakthrough is to posit a distinct, index-independent treatment of 'indexicals' (such as 'I', 'here' and 'now'), giving them a *character* (a function from context to content) and a contextually variable *content* (see Braun 2015 for a useful overview). For instance, 'I' for Kaplan has the character of referring to the speaker, and a content that varies depending on who is speaking. This requires two breaks from the Montagovian framework. The first is a posited intermediary between sentence and truth-value – *the proposition* – in which these contextually variable contents get deposited. (As Stalnaker (1970, 277) comments, propositions serve as 'an extra step on the road from sentences to truth values' which are 'justified only if the middlemen – the propositions – are of some independent interest ... ')

The second break from Montague is the deletion of certain index parameters – such as the agent and addressee parameters, which had been included to evaluate 'I' and 'you' – since their work is done through the effect of context on the proposition, prior to truth evaluation. (Here is the first step towards pulling information from the index into the proposition, and a crucial moral: to pull a given bit of information from the index into the proposition, find an associated indexical in the sentence to carry the information. As I claim in §§3–4, this is how to handle time and world information.)

Kaplan's picture of the semantic machinery (cf. Lewis 1980, 96) – which still remains orthodox – is thus:

So Kaplan posits a 'double dependence' of truth on context: context can operate on the sentence, varying the proposition expressed (indexicals); but context can also operate on the index, varying the settings of the parameters the proposition is evaluated against (world and time).

It is worth flagging some further Kaplanian assumptions, which I will adopt but cannot defend here: the propositions expressed by sentences at contexts are *structured contents* with *constituents* (Kaplan 1989a, 494), some of these constituents are determined by *variables* which may occur *free* or *bound*, and the free variables get their contents through a *variable assignment*, viewed as a component of *context* (Kaplan 1989b, 591). (This treatment of the variable assignment is controversial in ways that will prove directly relevant. I return to the matter in §§4–5.)

For present purposes, the key question is why Kaplan keeps the index in the picture, rather than trying to handle all context-dependence via the effect of context on the proposition.[3] Part of the reason is that Kaplan regards simple sentences like 1 as world and time neutral (§1), and so sees no world and time indexicals to exploit. (It bears noting that extensional treatments of world reference were barely on the map for Kaplan, and that intensional treatments of world and time reference were

---

[3] I am indebted to Herman Cappelen and to Brian Rabern for helpful discussions on this question.

common ground to Montague and Kaplan. So our key question was barely at issue for Kaplan.)

But Kaplan and also Lewis (1980) do offer two explicit arguments for keeping the index in the picture. Kaplan (1989a, 502–503; see also Cappelen and Hawthorne 2009, 70–73; and Schaffer 2012, 145–146) primarily offers *the operator argument*, which proceeds from the assumption that temporals and modals (like 'eternally' and 'necessarily') are operators, taking in a neutral sentence and holding it up to different values of an associated index parameter. Such a treatment indeed requires an index. As Kaplan (1989a, 503) explains:

> If we built the time of evaluation into the contents (thus removing time from the circumstances leaving only, say, a possible world history, and making contents *specific* as time), it would make no sense to have temporal operators. To put the point another way, if *what is said* is thought of as incorporating reference to a specific time, or state of the world, or whatever, it is otiose to ask whether what is said would have been true at another time, in another state of the world, or whatever. Temporal operators applied to eternal sentences (those whose contents incorporate a specific time of evaluation) are redundant. Any intentional operators applies to *perfect* sentences (those whose contents incorporate specific values for all features of circumstances) are redundant.

But what is not obvious is why Kaplan assumes (as embedded in his formalism, 545) that temporals and modals are operators in the first place. (In §§3–4 I sketch an extensionalist alternative which treats temporals and modals as quantifiers rather than operators. What is not obvious is what motivates preferring the intensional package of indices and operators, over the extensional package of variables and quantifiers.)

Kaplan (1989a, 508–510; see also Lewis 1980, 85–88 and King 2003, 201–206) offers a related concern, as to the relation between:

2. I am here now
3. Necessarily I am here now

As he notes, 2 is something like a logical truth: 'one need only understand the meaning of [2] to know that it cannot be uttered falsely'. But 3 can (and in many contexts does) express a falsehood. The puzzle is that 3 looks to follow from 2 by Necessitation. His solution is to say that the logicality of 2 stems from it being true at any proper speech context, but that the falsity of 3 is allowed by letting 'Necessarily' shift the world parameter to 'improper' settings on which the agent is not located at that time and place in that world. Thus he (1989a, 509) concludes:

> The difficulty, here, is the attempt to assimilate the role of a *context* to that of a *circumstance*. The indices < w, x, p, t > that represent contexts must be proper in order that ['I am here now'] be a truth of the logic of indexicals, but the indices that represent circumstance must include improper ones in order that ['Necessarily, I am here now'] not be a logical truth.

Relatedly, Lewis (1980, 86) says that indices are needed for independent one-feature-only variation under operators, to reach improper settings: 'The proper treatment of shiftiness requires not contexts but *indices*: packages of features of context so combined that they *can* vary independently.' He (1980, 88) concludes:

> To do their second job of helping to determine the semantic values of sentences with a given sentence as a constituent, the semantic values of sentences must provide information about the dependence of truth on indices. Dependence on contexts won't do, since we must look at the variation of truth value under shifts of one feature only. Contexts are no substitute for indices because contexts are not amenable to shifting. Contexts and indices will not do each other's work. Therefore we need both.

But Lewis (1980, 88–90) does see an alternative – the 'schmentencite way out' – which dispenses with index-dependence by arguing that it 'was needed only for the treatment of shiftiness, and we might claim that there is no such thing'. Indeed. Lewis's schmentencite dispenses with indices by having it that sentences never embed under operators, so shiftiness of sentence under operator never occurs – what occurs as constituents embedded under temporals and modals are not sentences but *schmentences*. So if the semantic value at a given context of 2 is *p*, the schmentencite denies that the semantic value of 3 is *p*; and if the semantic value at a given context of 1 is *q*, she denies that the semantic value of the following is given by placing a 'temporal box' in front of *q*:

  4. *I am eternally writing*

She thereby denies that 3 follows from 2 by Necessitation, and allows that 1 might be a true eternal sentence but 4 still a falsehood. The content of 1 (at a given context) need not be a time-neutral thing for temporal operators to shift, if 4 does not even have operator-sentence structure.

Lewis's response (see opening quote) manages to be both fully concessive and utterly contemptuous, granting the schmentencite her 'victory' but calling it 'cheap and pointless'. My best guess is that Lewis thinks of the schmentencite view as formally possible but completely ad hoc, and sees no independent motivation for treating the embedded sentence-like material as non-sentential save to block the operator argument.

(My schmentencite finds independent motivation for treating embedded sentence-like materials as *open formulae*, and so she re-claims her victory but imbues it with purpose.)

It is worth continuing to note (from the end of §1) that Kaplan and Lewis are largely taking up the Montagovian image of natural language as an intensional system, still assuming that simple sentences like 1 are world and time neutral, and still offering few constraints on index parameters (§§5–6).[4] Indeed Lewis (1980, 99) concludes by acknowledging that Kamp, Cresswell, and others have shown that 'there is a measure of disappointment in store' since empirical adequacy requires that 'we will have to repeat the world or time coordinates of our indices as many times over as needed' – that is, infinitely many times (§§3–4).

This is the story of a motivated completion of the schmentencite program (thus 'confessions of a schmentencite'). It is also the story of a kind of hyper-Kaplanian program, extending Kaplan's treatment of indexicals all the way through to world and time reference.

3. The rise of eternalism, and the fall of the operator argument

While Kaplan's Context-Index Semantics remains orthodox, there is debate over how best to model time reference within his framework. Kaplan is a *temporalist*, positing time-neutral propositions that can take different truth-values relative to different evaluation times. But the dominant view nowadays – says Richard (2015, 39) – is that of *eternalists*, who posit time-specific propositions with fixed truth-values, and so replace implicit temporal reference through an index parameter with explicit time variables in the proposition.[5]

Thus Stalnaker (1970, 1978) preserves the background context-index machinery but pares Kaplan's $<w, t>$ indices down to $<w>$.[6] (Here is a second step towards pulling information out of the index and into the proposition, leaving us one step away from explicit semantics – only the pesky world parameter remains.)

For present purposes the key issues are *why* eternalists opt to move the reference to times over to the extensional machinery, and *how* they then

---

[4]Lewis (1980, 98) says that Context-Index Semantics has 'recourse to genuine context-dependence' and thereby can claim the virtue of 'shirk[ing] the quest for rich indices.' This is good, but Lewis still offers no constraints on which parameters one can posit. He merely gives up a constraint – managing all context-dependence – that looked to be forcing infinitely many parameters.

[5]Contemporary temporalists include Ludlow (2001), MacFarlane (2003), Recanati (2007) and Brogaard (2012). Contemporary eternalists include Richard (1981), Salmon (2003), King (2003), Stanley (2005), Glanzberg (2009) and Soames (2011).

[6]Stalnaker (1970, 289) is actually non-committal on this score, allowing that there may be reason to add a time parameter to handle some tensed sentences. For present purposes I am interested in the move to drop the time parameter, and wish to credit Stalnaker with this idea, even if he was not fully committed to it.

answer the Kaplan–Lewis operator argument. As I see it, there are two primary rationales for eternalism, which are connected to the two original rationales for separating individual from world and time reference (§1). Basically, it was seen that these two rationales for separation fail for time reference, and hence that individual and time reference work in a parallel way, and so are appropriately handled by parallel formalisms. (All that remains is to extend this case to world reference).

So the first guiding rationale for separating time and individual reference was the thought that simple sentences like 1 ('I am writing') are time neutral. But subsequent work on tense undermines this naive image of time neutrality. So one main line of argument for eternalism – tracing back to Partee 1973 (see also Partee 1984 and Kratzer 1998; see Schaffer (2012, 135–136) for an overview of some of the relevant data) – is that tense displays all of the features of pronominal reference (a paradigm of explicit reference), displaying bound readings, strict/sloppy ambiguities, and E-type readings, *inter alia*, all of which look to be characteristic behavior of variables. Thus Partee (1973, 601) concludes: '[T]he tenses have a range of uses that parallels that of the pronouns, including a contrast between deictic (demonstrative) and anaphoric use, and ... this range of uses argues in favor of representing the tenses in terms of variables ... '

The referential view of tense thus undermines Kaplan's (1989a, 503; Kamp 1971, 231) idea that simple sentences like 1 contain 'no *temporal* indexical'. And so Enç (1986, 421; see also Ogihara 1996; Kusumoto 2005) says that, since time information is explicit, one should 'abandon the notion that intensions are functions from times and worlds, and maintain perhaps that they are only functions from possible worlds'. And likewise, King (2003, 223) maintains: '[I]f the proper way to treat tenses is *not* as index shifting sentence operators, then there is no need for temporal coordinates in indices of evaluation.' These theorists are following the Kaplanian playbook (§2): to pull a given bit of information from the index into the proposition, find an associated indexical in the sentence.[7]

---

[7]This is not to say that using explicit time variables *rules out* an intensional treatment. As Ninan (2012, 403) clarifies, one can have explicit time (/world/individual) variables and treat temporals (modals/quantifiers) as quantifiers, but also tack on obligatory wide-scope time (/world/individual) lambda binders so that these variables cannot appear free. The resulting semantic values are time (/world/individual) neutral, and so require an index with a time (world/individual) point to compute a truth-value. I agree with Ninan that such a view is consistent, though I see no empirical motivation for positing such obligatory wide-scope lambda binders. King (2003, 221; see also Glanzberg 2011, 119) takes the view that the extensional treatment is still a 'simpler, more elegant, less ad hoc treatment.' I agree with King, so long as the extensional treatment is also extended to worlds (*pace* King 2003, 28–29; though see Schaffer 2012, §4.1), for otherwise the intensional treatment of time can still claim the advantage of preserving the world-time parallel.

This idea of representing tense in terms of explicit time variables is naturally connected with developments in syntax. Current views of syntax generally posit a *small clause* where the root verb meets its arguments, with obligatory *mood* and *tense* projections (and other projections such as aspect which are not relevant here):

[$_{MP}$ Mood … [$_{TP}$ Tense … [$_{VP}$ Small Clause]]].

The small clause is not a stand-alone matrix clause but requires inflectional elements including mood and tense (see for instance Cinque 1999; see Glanzberg 2011, 116–120 for a summary of philosophically relevant issues). Thus in 1, 'am writing' is not a single element but rather how English spells out the blend of a root verb ($\sqrt{write}$) with indicative mood, present tense and other elements. The mood phrase is a natural repository for world information and the tense phrase is a natural repository for time information. There are more things specified in the sentence than are dreamt of in Index Semantics, and in particular, there is a place for explicit world and time indexicals. (Though the case of worlds must wait til §4.)

The second guiding rationale for separating time and individual reference was the thought that natural languages are expressively impoverished vis-a-vis times, only storing a single time perspective. But it soon emerged – from the work of Kamp (1971) and Vlach (1973) – that natural language is not expressively impoverished whatsoever vis-a-vis times, and in fact has the expressive power of an extensional system capable of 'storing' indefinitely many temporal points for further reference, just as an infinite stock of time variables would ensure. So a second main line of argument for eternalism surfaces, on which time reference in natural language has exactly the expressive power predicted by an extensional system of explicit variables.

This is not to say that the intensionalist treatment is refuted. Indeed it is known that, for any extensionalist treatment, there is a logically equivalent intensionalist reformulation, deploying infinitely many parameters and a fitting repertoire of operations to shift amongst them (Quine 1960; see also Kuhn 1980 and Cresswell 1990). Rather the point is that an extensional system immediately *predicts* the expressive power of natural language, while the intensional system needs epicycles just to *mimic* this successful prediction. In this vein van Benthem (1977, 426), contrasting the tense logic tradition with 'the use of predicate-logical formulas containing moment variables', notes:

[I]f one is willing to increase the complexity of the index to any extent (while adding enough operators to take profit of it), there is no need ever to resort to predicate logic *technically*, but in our opinion it is a Pyrrhic victory.

The deeper point is that the two guiding rationales for separate formal treatments of individual and time reference have both collapsed, and instead, the empirical phenomena of individual and time reference have proven relevantly similar, with exactly the referential features and expressive power that the explicit program predicted from the start.

Obviously, I cannot hope to settle the case for eternalism here. But for present purposes, I only need the less controversial claim that the eternalist position with explicit time variables is empirically open.

What becomes, however, of the Kaplan–Lewis operator argument for temporalism (§2)? I sketch a natural eternalist reply, that begins with the claim that the content of simple sentences like 1 is determined through a free time variable, which the assignment of the context sets to the speech time. So if Sadie is speaking at noon on 26 March 2018, then (irrelevant details aside) the essential content expressed at this context is[8]:

Sadie writes at noon on 3/26/18

This makes tokens of 1 talk about the speech time, and also ensures the redundancy-in-any-context of 1 with:

5. I am now writing

On this treatment, 5 merely specifies the speech time twice, both through the indexical 'now' and through the free variable set by the assignment of the context to the speech time. So far, so good.

Secondly, temporals like 'eternally' are regarded as quantifiers not operators. (Operators essentially mimic the effect of quantifiers in cases where it was thought that there is no explicit variable to bind.) In particular, 'eternally' is a universal quantifier binding free time variables in its scope. This makes 4 go false, with the essential content:

($\forall t$) Sadie writes at t

Thus the temporal quantifier binds a hitherto free variable.[9] This shows how temporals are not redundant. 1 can be true but 4 still false, as was wanted.

---

[8]Here and below I report structured contents via a casual blend of English and logic, to convey claims about constituency (e.g. that the time at issue is a constituent of the content) in a simple and neutral way, without taking a stand on the canonical format for structured contents. See Soames (1987) and King (2007) for some detailed regimentations, either of which could be adopted here.

What I just described as a natural eternalist reply is equally a form of Lewis's 'schmentencite way out', on which the content of 1 at a given context is not a constituent of the content of 4 (§2). The unembedded 1 is treated as having a time-specific content (e.g. being about 3/26/18). But the content of the embedded material in 4 is not this time-specific content, as the hitherto free time variable in 1 now occurs bound rather than free. So in 4 we do *not* see the time-specific content of 1 getting embedded. *That* treatment would yield vacuously quantified contents such as:

($\forall t$) *Sadie writes at noon on 3/26/18*

What embeds is rather an open formula. Thus my eternalist take the schmentencite way out by denying that the content of 1 is a constituent of the content of 4 (at any context), and more specifically holding that the content of 1 is the content of a *closed sentence*, while the corresponding constituent of the content of 4 is the content of an *open schmentence*.

Indeed my natural eternalist reply just is Lewis's (1980, 89) third version of the schmentencite view:

> [W]e might decorate the schmentences with free variables as appropriate. Then we might parse 'There have been dogs' as the result of applying 'It has been that … ' to the schmentence 'there are dogs at *t*' where '*t*' is regarded as a variable over times.

My schmentencite says this and adds that 'It has been that … ' has the effect of adding a quantifier binding the hitherto free time variable.

Lewis (1980, 89) continues: 'Schmentences would be akin to the open formulas that figure in the standard treatment of quantification.' And indeed this eternalist/schmentencite treatment is parallel to the orthodox treatment of the unembedded sentence:

6. *He is dancing*

versus the bound counterpart:

7. *Everyone is dancing*

The content of 6 is determined via a free individual variable, which the assignment function sets to a contextually salient male. So if the contextually salient male is Irving, then (irrelevant details aside) the essential content expressed at this context is:

---

[9]This fits what Ninan (2012, 403) calls 'standard views of this sort' on which 'each VP has an argument place for a silent temporal pronoun which can either be bound or get its value from the variable assignment'.

Irving dances

But on the bound reading, 7 does *not* express the vacuously quantified claim embedding this sentential content, which would be:

(∀x) *Irving dances* ('Everyone is such that Irving dances')

Rather on the bound reading, 7 expresses a content embedding an open schmentence content:

(∀x) *x dances*

With the schmentencite way out of the operator argument open, the primary Kaplan–Lewis argument for an intensional treatment of time reference no longer compels. This is not to say that there are no other arguments to consider.[10] But it is to say that the primary motivation that originally drove theorists to use intensional machinery for time reference fails.

It is worth continuing to note (§1) that eternalists are still upholding the Montagovian image of natural language as an intensional system for worlds, still assuming that sentences like 1 are world neutral (though not time neutral), and still offering few constraints on index parameters. But something else has gone badly: the deep parallels between world and time reference – preserved in Montague and in Kaplan – have now been broken. Times have been shunted over to the extensional machinery of explicit variables, but worlds stand alone, the last to be left to the intensional machinery of implicit parameters.

4. Cresswell and Stone on worlds, and the pull of the world-time parallel

It was left to Cresswell (1990) – and also Stone (1997), Schlenker (2006), Schaffer (2012) and Cresswell and Rini (2012) – to argue that the considerations arising with time reference arise in parallel ways with world reference. Just as the eternalist posits time-specific propositions, so the *necessitarian* posits world-specific propositions (as opposed to the *contingentist*, who upholds the image of world-neutral contents evaluated through an index featuring world parameters). She posits explicit world variables in the content. Just as the eternalist says that what is said in 1 is about the present time, so the necessitarian says that what is said in 1

---

[10]For example, Recanati (2007) offers an argument from cognitive and language learning considerations, and Brogaard (2012) offers an argument from claims about perceptual content. I cannot review these or other arguments in this space.

is about the actual world. Just as the eternalist contrasts 1 with past tense variants about past times such as:

8. I was writing

So the necessitarian also contrasts 8 with subjunctive mood variants about hypothetical scenarios such as:

9. I would be writing

In a context in which the speaker is imagining what she would be doing if she did not have to grade papers, 9 is naturally read as being about this imagined scenario.

The necessitarian promises to complete the program of explicit semantics from Montague on to Kaplan on to Stalnaker and the eternalists, by deleting that last pesky world parameter <w>, and finally pulling both world and time information from the index into the proposition, leaving the index empty.

Necessitarianism is decidedly a minority view. The majority view is eternalism-contingentism, with time-specific but world-neutral propositions held against a world parameter for truth evaluation. But I think that the eternalist-contingentist combination lacks consistent motivation, since exactly the same arguments for eternalism can be mimicked in the modal domain as arguments for necessitarianism (and equally: the obvious objections to necessitarianism can be mimicked in the temporal domain as objections to eternalism, and the eternalist rejoinders can be mimicked). In general, time and world reference are relevantly similar.

So consider the two main arguments for eternalism above (§3). The first argument undermined the guiding temporalist rationale of time-neutral contents, by revealing time variables (supported by referential features, and the syntactic posit of obligatory tense phrases). But parallel considerations arise with worlds. Thus Stone (1997; discussed also in Schaffer 2012) convincingly emulates the Partee (1973) data, showing that not just tense but mood as well displays the full referential signature of pronominal reference. As Stone (1997, 7) summarizes:

> [T]he interpretation of modals offers the same range of effects that characterize the interpretation of pronouns and tense. The only difference is the type of object involved. Where pronouns refer to individuals, tenses refer to times/events, and modals refer to hypothetical scenarios.

Percus (2000) argues for explicit world pronouns, and Speas (2004, 266) maintains: 'The evidence for a world argument comes from the fact that the world within which a sentence is to be interpreted shows the same locality conditions and restrictions on interpretation that pronouns and tense do.' And Schlenker (2006, 504), drawing these considerations together, speaks of 'a pervasive symmetry between the linguistic means with which we refer to [individuals, times and possible worlds]', and thus (2006, 509) calls for 'treating times and possible worlds in the way that we treated individuals' (2006, 509).

And syntactically – as already discussed in §3 – the tense projection comes alongside an obligatory mood projection, which is a natural repository for world information. In just the way that a tense projection specifies the time at issue, a mood projection specifies the world at issue. Thus the idea of world-neutral contents is just as syntactically naive as the idea of time-neutral contents.

The second argument for eternalism undermined the temporalist rationale of expressive poverty, by revealing that natural languages have the expressive power that the extensional system predicts for times. But the same considerations arise with worlds. Thus Cresswell (1990, 34–46) shows that natural language has the expressive power of an extensional system capable of 'storing' indefinitely many world points, just as an infinite stock of world variables would ensure, concluding (1990, 45): '[A]ll sentences are to be evaluated at a denumerably infinite sequence of worlds in a manner exactly analogous to the treatment of the temporal case ... ' As Kratzer (2014, §5) aptly summarizes, drawing the 'unified machinery' conclusion:

> Cresswell 1990 presented parallel arguments for modal anaphora, and showed more generally that natural languages have the full expressive power of object language quantification over worlds and times. Quantification over worlds or times is thus no different from quantification over individuals, and should be accounted for in the same way.

As she explains, 'Natural languages have syntactically represented individual variables' and 'It would be surprising if [natural languages] used two different equally powerful quantification mechanisms.' There is neither need nor motivation for kludging together extensional and intensional machineries.

Overall, both Schaffer (2012) and Cresswell and Rini (2012) argue for semantic parallelism concerning world and time reference.[11] As Cresswell and Rini (2012, vix) say: 'If you are faced with an argument in the

---

[11]Though Cresswell 1990 goes for the opposite pole, opting for an implicit semantics across the board. He keeps individual, world, and time reference in parallel by shifting them all over to the intensional

philosophy of modality, there ought to be a corresponding argument in the philosophy of time which has the same structure.' This semantic parallelism should not be surprising given the deep parallels in our conceptions of time and possibility – what is surprising is that the eternalist-contingentist majority would break the parallel without motivation.

Obviously, I cannot hope to settle the case for necessitarianism here. But for present purposes, I only need the less controversial claim that the necessitarian position with explicit world variables is empirically open too (alongside the eternalist position with explicit time variables).

What then becomes of the Kaplan–Lewis operator argument (§2) for contingentism? It should come as no surprise that parallel maneuvers are open to the necessitarian as were open to the eternalist (§3). So a natural necessitarian approach – paralleling the eternalist/schmentencite approach of §3 – involves two elements. First, simple sentences like 1 are regarded as having a world-specific content, contextually determined through a free world variable, which the assignment of the context sets to the speech word. So if Sadie is speaking in *w17*, then the essential content expressed is:

Sadie writes in w17

This makes tokens of 1 about the speech world, and also ensures the redundancy-in-any-context of 1 with:

10. *I am actually writing*

On this treatment, 10 merely specifies the speech world twice, both through the indexical term 'actually' and through the free variable set to the speech world.

Secondly, modals like 'necessarily' are no longer regarded as operators but just as quantifiers. In particular, 'necessarily' is a universal quantifier binding free world variables in its scope. This allows the following to go false:

11. *I am necessarily writing*

Since 11, in any context with Sadie speaking, is treated essentially as expressing:

($\forall w$) *Sadie writes in w*

This shows how modals are not redundant, in that 1 can be true but 11 false. And it shows how 3 is not derivable by 2 by Necessitation, since 2

---

machinery. Cresswell and Rini 2012 (esp. ch. 9) opt for a more deflationary 'no fact of the matter' view. My reasons for preferring explicit semantics over implicit semantics are given in §6.

features a free world variable that occurs bound within 3. Thus Glanzberg (2011, 118) aptly notes:

> It is not clear whether the kind of sententiality needed by the operator argument can ever be found in the syntax of natural language. It does not appear to hold for modality either. Modal auxiliaries, verbal mood, etc. all live outside of VP, and appear to occupy heads around the T level. Like tense, they do not function syntactically as sentential operators.

Again this is the schmentencite way out. The unembedded 1 has the contextually determined content of a closed sentence, with a world-specific content determined by a free occurrence of a world variable evaluated from the assignment function. This content is not what embeds under the modal in 9. *That* would be the vacuously quantified content:

($\forall w$) Sadie writes in w17 ('Every world is such that Sadie writes in w17')

Thus the schmentencite denies that the contextually determined content of 1 is a constituent of that of 11, just as she denies that the contextually determined content of 1 is a constituent of that of 4 (§3). The semantic value of the embedded material is not that of a closed sentence but rather that of an open schementence. Again this is not an ad hoc maneuver purely aimed to cut out the index, but rather a maneuver that is independently motivated, and matches the standard treatment of individual reference (§3).

Once worlds are pulled from the index into the proposition, the proposition is at last rendered as a complete thought. The index is empty and may, at last, be discarded (or at least, Kaplan's core cases of time and the world turn out to require no indices. Further proposals for parameters such as location and perspective still need to be considered. But the background dialectic shifts: what were relatively lightweight proposals to add parameters to a pre-existing index would now bear the much heavier burden of justifying complicating the semantic machinery with an index at all. The price of positing parameters just went up).

There is one last wrinkle: some – such as Rabern (2013) – will say that the index has not been emptied because the *variable assignment* remains.[12] But, as made explicit in §2, I am following Kaplan (1989b, 591; see also Heim and Kratzer 1998, 242–244) in treating the assignment as a feature of the context, not the index. After all, the assignment is what makes sentences like 6 ('He is dancing') get contents like *Irving dances*, and

---

[12] My thanks to Jeff King and Brian Rabern for helpful discussions.

so looks like a way of delivering the speaker's contextually intended referent.

But it is not obvious that the assignment can be treated as a feature of context. Lewis (1970; cf. Lewis 1980, 90) treats the assignment as an index parameter.[13] There is a tension lurking: in the Kaplanian framework, determinants of what is said are supposed to be features of context, while shiftable operands are supposed to be parameters of indices. *But the assignment is both.* It both determines what is said when a variable is free, and serves as a shiftable operand of lambda binders and quantifiers (on a Tarskian treatment).[14] So it may be said that the Kaplanian framework I have been starting with comes broken, in ways that matter for how I would adjust it.

So Rabern (2013, 399) develops an alternative view on which the assignment is not a feature of the context but instead lingers as the one and only parameter of the index, with the semantic values of sentences at contexts not being propositions (type t) but rather being functions from assignments to propositions (type $<\gamma, t>$). While this is not a matter I can discuss any further here, I will keep Rabern's position on the table for further discussion (§5), as a close cousin of explicit semantics which shares some of its advantages.

And so I arrive at the end of a long arc, starting from Montague's Index Semantics, through to Kaplan's Context-Index Semantics and the eternalist revision, and on to the fully explicit semantic program of Context Semantics. The last pesky world parameter is finally emptied into the proposition. In Context Semantics, propositions are *perfect* – they contain all information needed for truth evaluation, including world and time information. All context-dependence is mediated by indexicals, including temporal indexicals (specified by the tense phrase) and modal indexicals (specified by the mood phrase). The overall shape of the semantic machinery is simplified. No index remains:

Context Semantics

Sentence ⟶ Proposition ⟶ Truth Value

Context

---

[13]Indeed Lewis (1980, 89) draws on the idea that the assignment is an index parameter, to say that the schmentences are still index-dependent: 'Truth of a schmentence at an index would be like satisfaction of a formula by an assignment of values to variables.'

[14]Tarski (1983) offers a recursive definition of when an assignment satisfies a formula, including the following clause for universally quantified formulae: Assignment a satisfies '$(\forall x)\Phi$' iff: for any assignment β that is an x-variant of a, β satisfies 'Φ'. The relevant point is that Tarskian quantifiers shift the starting point assignment to look at alternative assignment points.

In my experience, some philosophers will argue about the details of time or (more probably) world reference. I cannot engage with these issues in this space (though see Schaffer 2012 as well as Kneer manuscript). And some philosophers will argue that index parameters are still needed for further matters such as locations and perspectives, or for the assignment itself. I cannot even begin to discuss theses cases here. (Though see Schaffer 2011 for more on perspective – there I argue that perspective is not an obligatory syntactic projection like world or time, but rather syntactically projected from specific terms such as 'tasty', as is explicit in constructions like 'tasty to Ann'.)

But some philosophers will be more open to thinking through the prospects for explicit semantics, and will allow that eternalism and necessitarianism are at the very least available options. These philosophers will be curious as to whether an explicit Context Semantics offers anything to gain, and may well let their decision on eternalism and/or necessitarianism, or Context Semantics generally, turn on what gets them to the best overall semantic framework. This strikes me as the most reasonable approach, so accordingly I turn to argue for the general preferability of a fully explicit semantics to any implicit alternative.

5. The joys of explicit semantics, as compared with mixed semantics

No confession is complete without the promise of salvation at the close. Accordingly, I conclude by reflecting on the joys of an explicit semantics as seen in Context Semantics (§4). What if natural language semantics could be done in a uniform way for individual, world and time reference, using only the extensional machinery of variables and quantifiers? I hope that by exhibiting the joys of explicit semantics, I might help draw those uncertain about the arguments for eternalism and necessitarianism (§§3–4), or those uncertain about positing implicit parameters for locations, perspectives or other less central cases, towards a general theoretical preference for the explicit approach.

From my perspective, intensional languages form an interesting class of artificial languages, worthy of logical study. For natural language semantics, they were a fruitful mistake.

There are two relevantly different comparisons for explicit semantics. The first comparison – pursued in this section – is with *mixed semantics*, which invokes some extensional machinery and some intensional machinery. This is the dominant view and my primary target, for kludging together two different formalisms without empirical motivation. The second comparison – postponed til §6 – is with *implicit semantics* (Cresswell 1990), which invokes intensional machinery across the board (even for individual reference), and so in some respects is the 'mirror image' of explicit semantics. This view is seldom taken very seriously, but I think it deserves more serious consideration since it can also lay claim to treating the parallel phenomena of individual, world, and time reference in parallel ways.

As compared to mixed semantics, I claim that explicit semantics enjoys three main advantages, the first and most obvious of which is greater simplicity and elegance. It should be visible that explicit semantics invokes simpler and more uniform semantic machinery:

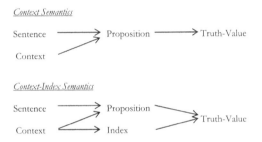

There may be reasons for complicating the formalism, but reasons must be given. The morale of §§1–4 is that the individual, world, and time reference prove to be relevantly similar. Natural language semantics is heir to two different formalisms, but these complications have outlived their rationale. (This is an invitation to the friend of mixed semantics to do better. What empirical phenomena compel such complications? What are the principled reasons for invoking different formalisms for these cases?)

As a second main advantage, explicit semantics is more constrained than mixed semantics, in multiple respects. The extensional machinery standardly comes with the following constraints baked in:

- the stock of variables posited is *infinite*
- the method of evaluating and shifting variables is through *the variable assignment* and *quantifiers*
- variables show up as components of *logical form*

In contrast, the intensional machinery predicts no comparable constraints. Rather the stock of points of reference is freely adjustable, the methods of evaluating and shifting these points are freely adjustable, and logical form provides no further guidance. (This is not to say that the intensional formalism is unacceptable, but only that the extensional formalism is preferable.)

Starting with the first bulleted point, the extensional machinery brings along an infinite stock of variables and a corresponding bold prediction about expressive power: natural language can track indefinitely many perspectives. It is uncontroversial that extensionalism yields such a prediction: indeed one of the original rationales for an intensionalist treatment of time and world was the thought that the extensionalist prediction was false, in that natural language could only track a single temporal and modal perspective (§§3–4).

But my point here is not just that the extensionalist prediction for time and world was eventually vindicated (§§3–4), but that the intensional machinery makes no comparable prediction whatsoever. Thus with time, temporalists went from positing one time parameter *t* (Prior, Montague, Kaplan), to two <*t1, t2*> (Kamp), to infinitely many <*t1, t2, ...* > (Vlach), just to fit the data. The number of temporal points posited is *a freely adjustable parameter*. This illustrates how the intensionalist machinery is less constrained. The temporalist semantics needs to be hand-tweaked to mimic what the eternalist semantics predicted all along.

Turning to the second bulleted point about evaluation and shifting, the extensional machinery brings along the assignment function and quantifiers. It should be admitted that the assignment function is not a very constrained posit. But we do have a strong independent grasp on the semantics of quantifiers (as seen in generalized quantifier theory: Barwise and Cooper 1981). So the extensionalist boldly predicts that temporals and modals both have the full range and structure of generalized quantifiers.[15]

But the intensional machinery does not carry along any fixed procedure for evaluating and shifting parameter values. With respect to evaluation, it is orthodox for parameters to be initialized from the context of utterance, so that the world and time at issue start off at the world and time of the

---

[15] Actually – following Bach, Kratzer and Partee (Partee 1995) – I take it that natural language quantification comes in two forms. There are *D-quantifiers*, modeled on determiners, in which syntax straightforwardly provides the [[Quantifier] [Restrictor]] [Scope] structure of generalized quantifiers. And there are *A-quantifiers*, modeled on adverbial quantifiers, which arguably lack a syntactically obligatory restrictor argument, and for which general discourse-level phenomena such as *the question under discussion* may play a key restrictive role. My specific prediction is that modals and temporals are A-quantifiers.

speech context (and only shift as a result of operators). But nothing mandates this procedure, and one of the insights of MacFarlane (2003) is that the intensional machinery is flexible. So MacFarlane not only adds a perspective parameter, but he initializes it from the circumstance of evaluation, to capture a form of relativist assessment sensitivity.[16] My point is that MacFarlane has revealed an independent respect in which index parameters are relatively less constrained, in that their initialization procedure is another freely adjustable matter for the intensionalist, which may even differ across different parameters.

With respect to shifting parameter values, there is no independent constraint on the number or structure of operations. Thus with time, temporalists went from positing simple 'Past', 'Future', 'Always Will', and 'Always Was' operators (Prior's P, F, G, and H), to Kamp's 'Since' and 'Until' augmentation, to all sorts of complicated things such as metric tense operators 'It was/will be the case *n* units of time before/after.' These developments proceeded – as van Benthem (1988, 10–11) notes – 'until a system arose whose temporal operators looked remarkably like the *Quine operators* for a variable-free predicate logic' so that 'the tense-logical formalism had become as strong as a full-blown, two-sorted predicate logic... freely allowing for quantification over [temporal items]'. Again the temporalist semantics needs to be hand-tweaked to mimic what the eternalist semantics predicted all along. Indeed while some mixed semanticists (Stanley 2005, 147–152) adopt the constraint of only positing parameters where natural language offers at least some associated operator, others (MacFarlane 2009, 244–246) reject even that constraint. So the range and structure of operators likewise seems to be a freely adjustable matter.

Turning to the third bulleted point about logical form, the extensional machinery posits variables and makes a correspondingly bold prediction about logical form: any information involved in truth evaluation is represented in logical form (with one admittedly worrisome exception: the assignment function). It is uncontroversial that extensionalism yields such a prediction: indeed one of the original rationales for an intensionalist treatment of time and world was the thought that the extensionalist prediction was false, in that simple natural language sentences such as 1 need time and world information for truth evaluation but do not represent the time or world at issue in logical form (§2).

---

[16]Of course I cannot discuss relativism here. But see MacFarlane 2003, 2005, 2014; Kölbel 2002; Egan, Hawthorne, and Weatherson 2005; and Lasersohn 2005 for key relativist proposals, and Schaffer 2011 and Stojanovic 2017 for some of the many critical discussions.

But my point here is not just that the extensionalist prediction proves defensible (via referential treatments of tense and mood: §§3–4), but that the intensional machinery yields no prediction at all. Indeed the entire point of the intensional machinery is to posit additional implicit points of reference thought to be *missing* from the proposition. So, in general, the number and variety of points of reference that the intensionalist posits float free from the logical form representation of any given sentence.

It is worth pausing to emphasize that logical form representations are subject to strong independent constraints. Some say that every element in the logical form is traceable to an element in the syntax (Stanley 2000, 401). This claim is controversial but it is at least widely accepted that syntax provides some starting point guidance.[17] Some add that variables in logical form may be diagnosed by various tests, such as binding (Partee 1989; Stanley 2000; Stanley and Szabó 2000; see Schaffer 2011 for this and other diagnostics). These tests are all controversial but it is at least widely accepted that there are decent (if fallible) diagnostics for variables. (The friend of mixed semantics can hardly object too vocally, since she too posits variables in logical form for those cases in which she uses the extensional machinery, and so she too needs to regard this as an overall well-constrained enterprise.)

No comparable constraints arise for parameters. Thus in §§1–2 I noted that theorists like Montague, Scott, Kaplan and Lewis freely tacked on parameters as needed. In this vein, Kaplan (1989a, 504) says: 'What sorts of intensional operators to admit seems to me largely a matter of language engineering. It is a question of which features of what we intuitively think of as possible circumstances can be sufficiently well defined and isolated.' This hardly sets a clear constraint. Lewis (1980, 84) says that the list of parameters 'should include time, place, world and (some aspects of) standards of precision' and adds: 'I am not sure what more should be added.' And MacFarlane (2009, 245–246) faces a 'proliferation' worry about 'opening the floodgates', which he puts as 'Pretty soon our nice ordered pairs will become ordered *n*-tuples!' His immediate reply is: 'Maybe you just need a lot of parameters to do semantics.' Perhaps. But it would be better to do semantics without this open-endedness.[18]

---

[17]In this vein, McKeever (2018) argues 'that one can reconcile the semantic criterion for positing variables with the demands of syntax' by positing covert syntax, and points out that positing covert syntax 'is pretty crucial to prevailing semantics methodology'.

[18]Stanley's (2000, 391) constraint that 'all truth-conditional effects of extra-linguistic context can be traced to logical form' is intended to block *free enrichment* (Sperber and Wilson 1986), whereby one can freely add constituents to the proposition. The complaint is that free enrichment lacks adequate constraints.

This concludes my discussion of the second main advantage of explicit semantics over mixed semantics, namely that explicit semantics provides a more constrained approach. I should emphasize that I am not claiming that mixed semantics is without any constraints, or otherwise theoretically unacceptable. I am only saying that explicit semantics induces stronger constraints (on the number of points of reference, their means of evaluation and shifting, and the implications for logical form), and is thus theoretically preferable, given that its bolder predictions are borne out.

A third benefit of explicit semantics over mixed semantics – and the final one I canvass – concerns the treatment of propositional truth that it engenders. A semantics without indices preserves the natural view that propositions are perfect and complete, which harkens back at least to Frege (1967, 38; also 1979, 135):

> A thought is not true at one time and false at another, but it is either true or false – *tertium non datur*. The false appearance that a thought can be true at one time and false at another arises from an incomplete expression. A complete proposition [*Satz*], or expression of a thought [*Gedankenausdruck*], must also contain the time datum.

But given implicit semantics, propositions are not complete but must be held up to further parameters. As Kaplan (1989a, 504) acknowledges: 'This functional notion of the content of a sentence in a context may not, because of the neutrality of content with respect to time and place, say, exactly correspond to the classical conception of a proposition.' Likewise, propositions are not true or false *simpliciter*, but only relative to parameter settings. The intensionalist trades the property of truth for the relation of truth-at-a-sequence.

How important is it to preserve the classical image of the complete proposition which is 'either true or false – *tertium non datur*'? For those inured to the relativized notion of truth-at-a-sequence, the point may seem rhetorical at best. But for others this is a core platitude of semantics, to be preserved if possible. In this vein, Cappelen and Hawthorne (2009, 1) start from a platitude they label *simplicity*, which includes: 'There are propositions and they instantiate the fundamental monadic properties of truth *simpliciter* and falsity *simpliciter*' (2009, 1). As Lewis himself (1970, 18) says, in a different context: 'Semantics with no treatment of truth-conditions is not semantics.' To which it may be added that the relation of truth-at-a-

---

But index parameters are just another way to freely enrich the eventual truth evaluation without representation in the logical form. Instead of freely enriching the proposition to factor in what one wants, one freely holds up the proposition to whatever added factors one wants.

sequence is not the same thing as the property of truth. My own view is that truth-at-a-sequence may be a passable substitute for the real thing, but that the real thing is still to be preferred if available.

So I conclude that explicit semantics is not just more simple and elegant than mixed semantics, but that it adds welcome constraints, and delivers classical propositions able to bear truth-values. To those philosophers curious if an explicit Context Semantics offers anything to gain over mixed Context-Index Semantics, and who will let their decisions on eternalism, necessitarianism and Context Semantics generally turn on what yields the best overall semantic framework, I say: explicit semantics is the lovelier framework.

And what if the index still remains, as on Rabern's preferred view (§4)? This particular version of mixed semantics – call it *almost-explicit semantics* – operates with explicit time and world variables in a parallel way just like explicit semantics, and posits only a single well-constrained index parameter for the assignment. Almost-explicit semantics cannot erase the index or fully recover complete propositions, but it does come with all of the constraints that come with positing world and time variables in logical form, and not allowing index parameters to be freely posited, freely initialized, or freely operated on. Rather we have just one parameter, initialized from the context, and operated on by standard quantifiers. Not bad. I thus reserve almost-explicit semantics as a backup view.

6. The joys of explicit semantics, as compared with implicit semantics

I have argued that explicit semantics is preferable to mixed semantics (§5). It remains to compare explicit semantics to implicit semantics (Cresswell 1990), which invokes intensional machinery across the board, even for reference to individuals. Both explicit and implicit semantics (but not mixed semantics) can boast a uniform and elegant treatment of individual, time, and world reference. For this reason, I think that implicit semantics deserves more serious consideration than it usually receives.

That said, I claim that explicit semantics enjoys three main advantages over implicit semantics, the first of which is a far more natural treatment of individual reference. It is worth saying why implicit semantics is seldom taken very seriously: the extensional treatment of individual reference seems so natural. To return to 1 ('I am writing'), it looks as if the reference to the speaker is glaringly explicit. This is in contrast to reference to world and time, which are only subtly explicit (in the mood and tense features bundled in with the root verb). So it seems to me that – once the deep parallels between individual, time and world reference emerge – the most reasonable stance is to extend the clear case of explicit reference out from individuals to world and time, rather than backtracking and denying it for individuals. (Here is a natural route to explicit semantics: Start with the orthodox extensionalist treatment of individual reference. See that world and time reference turn out relevantly parallel. So extend the extensionalist treatment to world and time reference. What went wrong?)

The second advantage of explicit semantics over implicit semantics is that explicit semantics is more constrained. Though the comparison is more subtle than with mixed semantics (§5), since the friend of implicit semantics is not committed to positing an infinite stock of variables, an assignment function, or quantifiers in any case (since she avoids the extensional machinery altogether). So, for instance, she can run a reasonable *tu quoque* argument about the lack of constraints on the assignment function (which parallels the lack of constraints on parameter initialization).

But explicit semantics remains more constrained in terms of logical form (§5). Using extensional machinery commits one to the bold prediction that any bit of information involved in truth evaluation shows up as a component of the logical form (again with the worrisome exception of the assignment function). Using intensional machinery does not commit one to any comparably bold prediction. For instance, the question of whether any locational information is involved in truth evaluation, and if so whether it be a single location, two locations, or an infinite sequence of locations, is a question all of whose answers – for the intensionalist – involve the very same location-neutral logical forms, and seemingly can only be answered by something like Kaplan's open-ended enterprise of 'language engineering'.

The third advantage of explicit semantics over implicit semantics is that it preserves the classical Fregean image of complete propositions bearing absolute truth-values (§5). Again this is a point that some will dismiss as mainly rhetorical, but others will regard as very serious. And again I say that preserving platitudes about propositions and truth is at least a benefit, leaving open how major a benefit it is.

I do not think that any of the listed benefits of explicit semantics over either mixed semantics or implicit semantics are conclusive, and I am open to the prospect that explicit semantics has costs I have not contemplated. But for now, I must conclude that explicit semantics is a lovely framework, preferable to the alternatives. If Lewis's schmentencite can claim her victory at long last, it is neither cheap nor pointless. Rather it arises only after the deep semantic parallels between individual, time and world reference are shown, and only in the wider context of choosing the best overall semantic framework.[19]

Disclosure statement

No potential conflict of interest was reported by the author.

References

Barwise, John, and Robin Cooper. 1981. "Generalized Quantifiers and Natural Language." *Linguistics and Philosophy* 4: 159–219.
Benthem, Johan van. 1977. "Tense Logic and Standard Logic." *Logique et Analyse* 80: 395–437.
Benthem, Johan van. 1988. *A Manual of Intensional Logic*. 2nd ed. Stanford, CA: Center for the Study of Logic and Information.
Braun, David. 2015. "Indexicals." *Stanford Encyclopedia of Philosophy*. Accessed September 9, 2017. https://plato.stanford.edu/entries/indexicals/.
Brogaard, Berit. 2012. *Transient Truths: An Essay in the Metaphysics of Propositions*. Oxford: Oxford University Press.
Cappelen, Herman, and John Hawthorne. 2009. *Relativism and Monadic Truth*. Oxford: Oxford University Press.
Cinque, Guglielmo. 1999. *Adverbs and Functional Heads: A Cross-Linguistic Perspective*. Oxford: Oxford University Press.
Cresswell, M. J. 1972. "The World Is Everything That Is the Case." *Australasian Journal of Philosophy* 50: 1–13.
Cresswell, M. J. 1990. *Entities and Indices*. Dordrecht: Kluwer.
Cresswell, M. J., and A. A. Rini. 2012. *The World-Time Parallel: Tense and Modality in Logic and Metaphysics*. Cambridge: Cambridge University Press.
Egan, Andy, John Hawthorne, and Brian Weatherson. 2005. "Epistemic Modals in Context." In *Contextualism in Philosophy: Knowledge, Meaning, and Truth*, edited by Gerhard Preyer and Georg Peter, 131–168. Oxford: Oxford University Press.
Enç, Mürvet. 1986. "Toward a Referential Analysis of Temporal Expressions." *Linguistics and Philosophy* 9: 405–426.
Frege, Gottlob. 1967. *Kleine Schriften*. Hildesheim: Olms.
Frege, Gottlob. 1979. "Logic (1897)." In *Posthumous Writings*, edited by Hans Hermes, Friedrich Kambartel, and Friedrich Kaulbach, 126–151. Chicago, IL: University of Chicago Press.
Glanzberg, Michael. 2009. "Semantics and Truth Relative to a World." *Synthese* 166: 281–307.

---

[19]Thanks to Herman Cappelen, Delia Fara, Michael Glanzberg, Jeff King, Angelika Kratzer, Brian Rabern, Mark Richard, Roger Schwarzschild, Jason Stanley and Zoltán Gendler Szabó, the referees for *Inquiry*, and the audience at the Rutgers Semantics Workshop.

Glanzberg, Michael. 2011. "More on Operators and Tense." *Analysis* 71: 112–123.
Heim, Irene, and Angelika Kratzer. 1998. *Semantics in Generative Grammar*. Oxford: Blackwell.
Kamp, Hans. 1971. "The Formal Properties of 'Now'." *Theoria* 37: 227–274.
Kaplan, David. 1989a. "Demonstratives: An Essay on the Semantics, Logic, Metaphysics, and Epistemology of Demonstratives and Other Indexicals." In *Themes from Kaplan*, edited by Joseph Almog, John Perry, and Howard Wettstein, 481–563. Oxford: Oxford University Press.
Kaplan, David. 1989b. "Afterthoughts." In *Themes from Kaplan*, edited by Joseph Almog, John Perry, and Howard Wettstein, 565–614. Oxford: Oxford University Press.
King, Jeffrey. 2003. "Tense, Modality, and Semantic Values." *Philosophical Perspectives* 17: 195–245.
King, Jeffrey. 2007. *The Nature and Structure of Content*. Oxford: Oxford University Press.
Kneer, Markus. manuscript. "Semantic Incompleteness, Modal Anxiety, and Necessity."
Kölbel, Max. 2002. *Truth Without Objectivity*. London: Routledge.
Kratzer, Angelika. 1998. "More Structural Analogies Between Pronouns and Tenses." Proceedings of Salt VIII: 92–109. Cornell University Press.
Kratzer, Angelika. 2014. "Situations in Natural Language Semantics." *Stanford Encyclopedia of Philosophy*. Accessed September 28, 2017. http://plato.stanford.edu/entries/situations-semantics/.
Kuhn, Steven T. 1980. "Quantifiers as Modal Operators." *Studia Logica* 39: 145–158.
Kusumoto, Kiyomi. 2005. "On the Quantification Over Times in Natural Language." *Natural Language Semantics* 13: 317–357.
Lasersohn, Peter. 2005. "Context Dependence, Disagreement, and Predicates of Personal Taste." *Linguistics and Philosophy* 28: 643–686.
Lewis, David. 1970. "General Semantics." *Synthese* 22: 18–67.
Lewis, David. 1980. "Index, Context, and Content." In *Philosophy and Grammar*, edited by Stig Kanger and Sven Öhman, 79–100. Dordrecht: Reidel.
Ludlow, Peter. 2001. "Metaphysical Austerity and the Problems of Modal and Temporal Anaphora." *Philosophical Perspectives* 15: 211–227.
MacFarlane, John. 2003. "Future Contingents and Relative Truth." *Philosophical Quarterly* 53: 321–336.
MacFarlane, John. 2005. "Making Sense of Relative Truth." *Proceedings of the Aristotelian Society* 105: 321–339.
MacFarlane, John. 2009. "Nonindexical Contextualism." *Syntheses* 166: 231–250.
MacFarlane, John. 2014. *Assessment Sensitivity*. Oxford: Oxford University Press.
McKeever, Matthew. 2018. "Positing Covert Variables and the Quantifier Theory of Tense." Inquiry.
Montague. 1968. "Pragmatics." In *Contemporary Philosophy: A Survey, Volume 1*, edited by Raymond Klibansky, 102–122. Florence: La Nuova Italia Editrice.
Montague. 1970. "Universal Grammar." *Theoria* 36: 373–398.
Montague. 1973. "The Proper Treatment of Quantification in Ordinary English." In *Approaches to Natural Language*, edited by Patrick Suppes, Julius Moravcsik, and Jaakko Hintikka, 221–242. Dordrecht: Reidel.
Ninan, Dilip. 2012. "Propositions, Semantic Values, and Rigidity." *Philosophical Studies* 158: 401–413.

Ogihara, Toshiyuki. 1996. *Tense, Attitudes, and Scope*. Dordrecht: Kluwer.
Partee, Barbara. 1973. "Some Structural Analogies Between Tenses and Pronouns in English." *The Journal of Philosophy* 70: 601–610.
Partee, Barbara. 1984. "Nominal and Temporal Anaphora." *Linguistics and Philosophy* 7: 243–286.
Partee, Barbara. 1989. "Binding Implicit Variables in Quantified Contexts." *Proceedings of the Chicago Linguistics Society* 25: 342–365.
Partee, Barbara. 1995. "Quantificational Structures and Compositionality." In *Quantification in Natural Languages*, edited by Emmon Bach, Eloise Jelinek, Angelika Kratzer, and Barbara Partee, 541–601. Dordrecht: Kluwer.
Percus, Orin. 2000. "Constraints on Some Other Variables in Syntax." *Natural Language Semantics* 8: 173–229.
Prior, Arthur. 1957. *Time and Modality*. Oxford: Clarendon Press.
Quine, W. V. O. 1960. "Variables Explained Away." *Proceedings of the American Philosophical Society* 104: 343–347.
Rabern, Brian. 2013. "Monsters in Kaplan's Logic of Demonstratives." *Philosophical Studies* 164: 393–404.
Recanati, Francois. 2007. *Perspectival Thought: A Plea for (Moderate) Relativism*. Oxford: Oxford University Press.
Richard, Mark. 1981. "Temporalism and Eternalism." *Philosophical Studies* 39: 1–13.
Richard, Mark. 2015. "Temporalism and Eternalism Revisited." *Truth and Truth Bearers: Meaning in Context* 2: 38–57.
Salmon, Nathan. 2003. "Tense and Intension." In *Time, Tense, and Reference*, edited by Quentin Smith, and Aleksandar Jokic, 107–154. Cambridge, MA: MIT Press.
Schaffer, Jonathan. 2011. "Perspective in Taste Claims and Epistemic Modals." In *Epistemic Modality*, edited by Andy Egan, John Hawthorne, and Brian Weatherson, 179–226. Oxford: Oxford University Press.
Schaffer, Jonathan. 2012. "Necessitarian Propositions." *Synthese* 189: 119–162.
Schlenker, Philippe. 2006. "Ontological Symmetry in Language: A Brief Manifesto." *Mind and Language* 21: 504–539.
Scott, Dana. 1970. "Advice on Modal Logic." In *Philosophical Problems in Logic: Some Recent Developments*, edited by Karel Lambert, 141–173. Dordrecht: Reidel.
Soames, Scott. 1987. "Direct Reference, Propositional Attitudes, and Semantic Content." *Philosophical Topics* 15: 47–87.
Soames, Scott. 2011. "True At." *Analysis* 71: 124–133.
Speas, Margaret. 2004. "Evidential Paradigms, World Variables and Person Agreement Features." *Italian Journal of Linguistics* 16: 253–280.
Sperber, Dan, and Dierdre Wilson. 1986. *Relevance: Communication and Cognition*. Oxford: Basil Blackwell.
Stalnaker, Robert. 1970. "Pragmatics." *Synthese* 22: 272–289.
Stalnaker, Robert. 1978. "Assertion." *Syntax and Semantics* 9: 315–332.
Stanley, Jason. 2000. "Context and Logical Form." *Linguistics and Philosophy* 23: 391–434.
Stanley, Jason. 2005. *Knowledge and Practical Interests*. Oxford: Oxford University Press.
Stanley, Jason, and Zoltán Gendler Szabó. 2000. "On Quantifier Domain Restriction." *Mind and Language* 15: 219–261.

Stojanovic, Isidora. 2017. "Context and Disagreement." *Cadernos de Estudos Linguísticos* 59: 7–22.

Stone, Matthew. 1997. *The Anaphoric Parallel Between Modality and Tense*. Department of Computer & Information Science Technical Reports, University of Pennsylvania.

Tarski, Alfred. 1983. "The Concept of Truth in Formalized Languages." In *Logic, Semantics, Metamathematics; Papers from 1923 to 1938 (Second Edition)*, translated by J. H. Woodger, 152–278. Indianapolis, IN: Hackett.

Vlach, Frank. 1973. "'Now' and 'Then': A Formal Study in the Logic of Tense Anaphora." PhD diss., University of California Los Angeles.

Positing covert variables and the quantifier theory of tense

Matthew McKeever

ABSTRACT
A crucial issue in the debate about the correct treatment of natural language tense concerns covert variables: do we have reason to think there are any in the syntax, as the quantifier theorist maintains? If not, it seems we can quickly discount the quantifier theory from consideration, without even considering the data in its favour. And, indeed, there is a good reason to doubt that there are such variables: contemporary syntactic theory, notably, does not seem to posit them. I respond to this argument going from the premise that positing covert variables is illicit to the conclusion that quantifier theories of tense are false. I argue that the argument fails by suggesting a non-committal understanding of the process of positing covert variables. On this view, even if we are not doing fundamental syntax in so positing, nevertheless there is a reason to think we are doing something theoretically productive, because it seems like an important part of the practice of semantics and because it seems like semantics is a discipline which is making progress. The progress of semantics supports the methodology of positing variables at its heart; the quantifier theorist of tense, accordingly, need not worry about their syntactic commitments.

1. Introduction

Natural language tense and modal expressions seem to be devices of generality, and in particular, to be what I will call quantifier-like.[1] For an expression to be quantifier-like is for it to be modellable in terms of the basic quantifiers of predicate logic, the existential and universal quantifier.

Thus consider a sentence like

(1)   I saw the Godfather II in College

---
[1]Having once mentioned modality, I will talk in this paper solely about tense; some conclusions generalise, although one ought to be careful about lumping the two together, given their markedly different syntactic realisations.

This is true provided *there is some* time in the past (in particular some time lying in the span of time at which I went to college) at which I saw the film. In a similar vein

(2) I have never seen the Godfather III

is true provided at *all* times in the past, it is not the case that I saw the film. Such glosses can be modelled by a simple formal quantificational language as so, where the variable 't' ranges over times and where 'u' stands for the utterance time:

- $\exists t. t < u$ & College(t) and SeeGFII(t)
- $\forall t. t < u \rightarrow \neg$ SeeGFIII(t)

(In a more accurate treatment which would command greater empirical coverage as an attempt to model natural language, we would use generalised quantifiers here, but this will suffice for our initial purposes.)

What is the significance of this fact? Nothing immediately, but we can note that there are other sorts of expressions of natural language that are quantifier-like. Most noticeably, some quantified noun phrases are. Thus we have:

(3) Every pandas like bamboo
(4) Some chairs are comfortable

Representable as:

- $\forall x.$ Panda(x) $\rightarrow$ LikesBamboo(x)
- $\exists x.$ Chair(x) & Comfortable(x)

Moreover, as we will see, people like Partee (1973), Cresswell (1991) and Schlenker (2006), among others, have argued that the similarities go much further, and the basis of this similarity, we might think that our semantic theory should treat quantified noun phrases and tenses in the same way.

My aim in this paper is to obliquely defend the quantifier theory of tense by responding in detail to an objection to it. The objection goes roughly as so: the most well-known way of modelling quantified noun phrases in natural language, the one we teach to our students and presuppose in our research, is by means of (generalised) quantifiers and bindable elements like pronouns and traces. If tense is to get the same semantic treatment as quantified noun phrases, then it must be modelled by means of quantifiers and variables.

But, on the surface of it, it is at least dubious that tenses should be modelled, syntactically, in this way. If tenses are quantificational, we should expect a syntactic form for 1 very roughly as so:

- PAST$_t$ I see(t) the Godfather II and I'm in college(t)

where 't' is a temporal variable bound by the temporal quantifier 'PAST'. The problem is it is unclear that we have any reason to make such a posit. There is at least prima facie plausibility in the claim that pronouns are variable-like elements, and traces, although theoretical posits, are to some extent independently motivated by syntactic data (such as wh-movement). By contrast, the syntactic evidence for the existence of temporal variables is questionable. Most obviously, there are no pronoun-like items which function *overtly* as temporal variables (although see Schlenker's interesting discussion of 'then' in Schlenker 2006). Moreover, if there were temporal variables, there would be temporal quantifiers, and one behaviour we expect of quantifiers is to be iterable, but things do not go well when we try to iterate temporal expressions (for but one example of this, see Dowty 1982).

That is the objection to the quantification theory I want to consider in this paper: that there are no syntactic grounds for positing temporal quantifiers and variables. If this is so, then the analogy with quantified noun phrases is weak, and we should seek another analysis of tense, and not take its quantifier-like behaviour as an insight into its actual semantics. There is a tension between the superficial quantifier-like behaviour of tense and what syntax gives us.

I will attempt to dissolve this tension, and to show that there is, in fact, nothing to worry about: semanticists, and thus tense theorists, need not worry if their analyses do not square with what syntax tells us.

My response has two steps. The first notes that positing covert variables, or covert structure more generally, is fundamental to how working semanticists operate. If it were no good, then semantics would essentially be barren – as I will show, even simple objectual quantification, according to its textbook analysis in Heim and Kratzer (1998), would be no good.

But semantics is not no good – it appears as if formal semantics is a research programme which is making progress. In the past half-century or so, we have developed a range of technical tools and proposed theories using them sensitive to a lot of linguistic data. These twin constraints of empirical and technical adequacy are ones formal semantics shares with uncontroversially successful disciplines, and absent reason to the contrary, if we think the latter make progress, we should think the former does too. But then if one shares my sanguine assessment of formal semantics and accepts that positing covert material is a central part of its methodology, one should be inclined to accept that positing such material is licit: it is allowed because it is a feature of a research programme which is successful. And so the objection to the quantifier theory of tense is defanged.

That is all very well, one might think, but it does not address the main problem, which is that syntax gives us no reason to think that these things we posit are actually there. Otherwise put, these three claims seem to be in tension: positing covert variables is a central practice to semantics; positing covert variables is (syntactically) unmotivated; semantics is not barren.

However, I do not think they are in tension, rightly understood. In particular, they are not provided one changes the way one understands what it is to posit covert variables, and in particular provided one gives up on the claim that positing covert variables involves doing syntax. Although it may *seem* like that is what we are doing, when we write our trees in semantics classes and papers, it is not. We are doing something different, and so it is no objection that our posits do not chime with syntactic theory. The obvious and crucial question, though, is then: what *are* we doing when we posit covert variables?

Well, I do not know. But I do not think that matters. We study and develop theories about many things we do not know really know the essence of. At the risk of an overblown example, we really have no idea what is going on when we use the Schroedinger equation to describe the evolution of a quantum system, but use it we still do, because it works. People working at the foundations of physics can and do ask what *is* really going on, but for most people, even working physicists and engineers, it does not matter. If they can use it in their code or rely on it when building some tiny implements, that is enough for them. It is not really an objection to a physicist or the engineer using the equation to say that the quantum world is completely mysterious and they do not *really* know what they are talking about. Similarly, it is not an objection to the semanticist, I claim, to say that positing covert variables is mysterious, because it seems to work, in the sense that it seems to be an important part of the method of semantics, and semantics seems to make progress.

So that is my view: in positing covert structure, we are doing *something*, and we do not know what it is, but that does not matter. It seems to work: we develop testable theories constrained by linguistic data and formal principles (such as compositionality), we refine and refute those theories and develop new ones. Given that it seems to work, even if we do not know what we are doing, we should feel no qualms continuing to do it. The quantifier theorist of tense, accordingly, should be unworried by claims that their theory rests on indefensible theoretical posits.

The structure of the paper is as follows. In the next section, Section 2, I briefly review the operator vs. quantifier debate in the theory of tense and show why the data seems to support a quantifier view. I will point out how this requires covert variables and then go on to consider another theory which also requires such variables, namely Stanley and Szabo's well-known view of quantifier domain restriction. I will then, in Section 3, consider a challenge to Stanley and Szabo's view, from Collins (2007), according to which we have no syntactic grounds for positing covert domain restricting variables and show how it generalises to temporal variables. Section 4 is devoted to responding to this challenge by presenting in more detail the argument sketched above, and Section 5 considers some responses and objections.

2. The case for variables

In this section, I want to review the case for positing covert variables both in the debate about tense and in semantics more generally.

In order to do this, a bit of familiar history. I noted in the introduction that tense in English appears quantifier-like: at least at first glance, it seems like we can model tenses by means of the quantifiers of predicate logic. I suggested, somewhat quickly, that on that basis one might seek to give tense and natural language quantified noun phrases the same analysis.

However, this presentation somewhat ignored the history of the debate, which in fact went a different way: the quantifier analysis of tense was not the first one, and to properly understand the nature of the debate, it is necessary to see this.

So, first, note that just because something is quantifier-like, it does not mean that it must be analysed in terms of quantifiers and variables bound by them. The reason for this is that, according to the way I am using it, modal operators are themselves quantifier-like. That is to say, a formula like:

- p

Is well-glossed in terms of the first order:

- $\forall w.$ Accessible$(w) \rightarrow$ True(p,w)

Clearly, the modal operators of, formal example, basic modal logic are not the same sort of thing as the quantifiers of, for example, the first-order logic. They may be truth conditionally equivalent to sentences of quantificational logic, but they are not themselves quantifiers.

It is accordingly a live option that, in the face of quantifier-like behaviour of a given sort of expression, one analyses it as a modal operator. And this is precisely what we find in the work of Prior (e.g. Prior 1968).

Prior thought one could analyse tenses in terms of a modal logic with four operators. He added to quantificational logic the four operators P, F, H, G, which express, respectively, past existential quantification, future existential quantification, past universal quantification and future universal quantification. Thus consider

(5) I saw the Godfather II in College
(6) I have never seen Godfather III
(7) I will understand tensor calculus some day
(8) I will always remember this day

These could be translated into Prior's temporal logic (or rather an Anglicised version of it) as so:

- P (I see Godfather II and I am in college)
- H ($\neg$ I see Godfather III)
- F (I understand tensor calculus)
- G (I remember this day)

This seemed reasonable. If our choice were between a quantifier and an operator analysis of tense, then in light of the superficial difference between objectual quantifiers and tenses, one might think it is wise to opt for the latter.

Indeed, even those who realised the inadequacies of the Priorean theory – of which more immediately below – were impressed by this fact. Hans Kamp, whose seminal work on 'now' showed there must be more to the theory of tense than is permitted in a simple, even multimodal modal logic like the above, wrote

> I of course exclude the possibility of ... symbolizing the sentence by means of explicit quantification over moments .... Such symbolizations ... are a considerable departure from the actual form of the original sentences which they represent – which is unsatisfactory if we want to gain insight into the semantics of English. (Kamp 1971, fn1)

However, time went on and it became more and more evident that the operator theory had some serious empirical difficulties. Partee (1973) famously showed some similarities between tenses and pronouns, noting that just as we use pronouns to refer to some particular object, so we sometimes use tenses to refer to some particular time. To use a variant of her example, imagine we are on I-35 driving back to work, having just eaten lunch at home. I remark in a panic:

(9) I didn't turn the stove off!

The existential truth conditions given by the operator analysis do not seem right: it is quite tempting to say that we are talking about some particular time, namely the time just after I finished cooking dinner and asserting, of it, that I did not turn the stove off then.

More work in this vein continued, as a range of sentences appeared which were not immediately amenable to an operator analysis. We may think of Dowty's (1982) work on adverbs, which shows the operator analysis yields incorrect predictions if we analyse adverbs as operators, and later Cresswell (1991)'s work which – building on that of Kamp and Vlach (1973) – showed that there are sentences not analysable in terms of a modal logic anything like the one above. A particularly neat example is furnished by sentences involving interactions between tense and universal quantifiers. Consider:

(10) Some day, each of my students will be on the board of *Linguistics AndPhilosophy*

The most salient reading of this sentence is that at some one point in the future, each of the speaker's *current* students will be on the board of the journal – the speaker is imagining a situation in which his school rules supreme by capturing the editorial board. We cannot get this from the modal analysis though. It can give us the following disambiguations:

- F (Each of my students be on the board of *Linguistics and Philosophy*)
- [Each of my students]$_x$ F (x be on the board of *Linguistics and Philosophy*)

Neither of these give us the most salient reading. The first gets us that at some time in the future, each of the speaker's then students will be on the board. That is, it talks about the wrong students – the future ones, which are not necessarily the current ones. The second has it that each of the speaker's students are such that in the future he or she will be on the board – but it does not mean that all the students will be on the board *together*: one could be on the board at t_2 and only t_2, one at t_3 and only t_3 and so on. This does not secure the editorial board domination reading we are looking for.

By contrast – although I will not get into the details here – if one were to analyse tenses as quantifiers binding temporal variables, these problems would not arise. On the basis of this (and other data), it has come to seem that a quantificational theory is more apt for an empirically adequate treatment of tense, and that we should thus adopt it. As King says

> If the complex temporal facts present in natural language are most readily and easily represented by viewing tenses as involving explicit quantification over time and as expressing relations between times, that is a good reason for thinking that tenses really work this way. (King 2003, 218)

However, recall the Kamp point mentioned above – he felt uncomfortable with the quantificational analysis because it did not do justice to the surface form of the sentence. What has happened to that?

Well, it seems that positing covert stuff has become more acceptable in the intervening years. Indeed, if one looks around at practicing semanticists and philosophers of language, you will see a lot of positing syntax.[2] Just to pick more or less at random from my recent reading, we might think here of predicativism about names (defended most recently by Fara 2015), which involves positing a covert definite article which precedes bare names or the covert type-shifting operation appealed to in Moss (2015) to get her theory of epistemic modals to work.

---

[2] It is worth noting that syntax has also, as one might expect, moved on in the intervening years, a fact worrying given the standard semantics textbook is now 20 years old, and as such rests on 20 years old syntax. I thank a reviewer for this journal for pointing out this fact about the development of syntax.

Let me consider one more example, a very famous and influential one in philosophical circles, which I will later make use of. According to Stanley and Szabó (2000), we need to account for the well-known problem of covert domain restriction by positing covert variables in the syntax. Their argument for this claim is, in essence, to say that covert domain restriction is a quantificational phenomenon, and since quantification is handled in natural language semantics with variables and operators which bind them, we need such technical tools in this case.

The crucial observation here is that sometimes the implicit restrictions on the quantifier domain associated with an expression vary with the values of a higher quantifier which appears to control that expression. Here is an example. The head teacher of Eton, praising the performance of all his final year students, which are divided among several classes, can say:

(11) In every class, every student passed.

The first thing to note is that the first quantifier 'every class' is covertly restricted: we are not asserting anything about every class *in the whole world*, but rather about every class *in the final year at Eton* (even that needs to be further restricted to the school year when the teacher is speaking, but let us ignore that). However, if that were all there was to covert domain restriction, one might think a pragmatic analysis in terms of loose speech or something similar might suffice. But there is more: the second quantifier also is covertly restricted, but not once and for all. Rather, there are different restrictions relative to different classes. We are saying something like *in every class, every student in that class* passed. Less colloquially, one could rephrase that as *in every class x, every student in x passed*. That is to say, it seems like the domain restriction is something like *being in x* where *x* is a variable bound by 'every class' (or rather, technically speaking, by its associated lambda binder).

The truth conditions of the sentence, then, are equivalent to truth conditions which, when they make explicit the domain restriction associated with the lower quantifier, include a variable in the restricting phrase which is bound by the higher quantifier. Stanley and Szabo in effect argue that when this occurs, there is *in fact* a covert variable in the syntax which gets bound. Whenever we have the semantic effects of variable binding, we have its standard syntactical realisation as well.

They posit, accordingly, logical forms (LFs) very roughly like so:

(12) [$_s$[$_{pp}$ [In [$_{dp}$ every class Fy]$_x$], [$_s$[$_{dp}$ every student Gx] [$_{vp}$ passed]]]

If the property variable F gets assigned something like the property *comprising the final year at Eton*, the variable 'y' gets assigned nothing, and the variable G gets *being in x*, we will get the right results.

The details – which have been the subject of some debate – are not so important right now. The important point to note is that both we have here another case of positing covert variables, and also a rationale for doing so: we may posit them, Stanley and Szabo say, when we witness the semantic effects of covariation. If they were right, then granted we do witness the semantic effects of covariation in the tense case, we would be right to posit temporal variables to account for tense in natural language.

3. The case against covert variables

However, there are problems with this claim. Most pertinently for the current paper, the syntactic bona fides of these covert variables has been called into question.

In a sense, this worry has been in the debate from the start – it is just been overlooked. Recall the passage from Kamp quoted above – he did not consider quantificational approaches to tense because it certainly did not appear that the syntax supported them. Kamp (or the timeslice of him writing in 1971) would not be too moved by Stanley and Szabo's view that semantic binding implies syntactic binding. The question is: should he, and we, be moved?

Well, it seems like the way to answer this question is just to look and see what syntacticians say. Does syntax support Stanley and Szabo's posit? As it happens, a recent paper by Collins (2007) considers exactly this question.

The answer, however, is not positive. Collins shows that Stanley and Szabo do not succeed in making the case that there are covert variables. Collins' discussion is quite long and involves some syntactical technicalities which it would take us too off course to consider, so I will just pull from his discussion two points which I think are especially problematic to the friend of variables.

So consider: what would one need to do to make the case that there were variables in syntax? It seems, minimally, there are at least two things: one would have to say what sort of syntactic items these variables were and what sort of syntactic relations they enter into with the rest of the sentence. Collins thinks Stanley and Szabo fail on both these counts.

As to the first, he notes that a variable is not a syntactic notion. Positing variables in the syntax is something like a category mistake: it is like positing water in the syntax.[3] He points out that lexical items are bundles of features, as, for example, the pronoun 'he' encodes the features of third person, singular and male. But variables have no such features (Collins 2007, 832), and so one will fail to answer the question as to what sort of syntactic items they are.

As to the second point, can we make sense of the syntactic relations the variable complexes like '*Gx*' in the above form stand in to the rest of the sentence? Again, Collins suggests the answer is no. He points out that Stanley himself has wavered on this question, sometimes loosely suggesting that they 'co-habit' the nodes of the DP they are associated with. The problem is 'co-habitation' is not a syntactic idea, and so will not suffice. He also suggested they might function as adjuncts, but Collins shows this too is problematic. The basic thought is that we want a syntactic reason for positing these adjuncts. But Stanley seems to claim that two sentences can be syntactically the same yet one might permit and the other not permit the domain restricting adjunct. For example, in a reply to a paper by Cappelen and Lepore (2002), Stanley argues that the first sentence below contains covert location variables whereas the latter does not:

(13) Everywhere I go, it rains.
(14) Every I go, 2+2=4.

The thought is we need them in the former case because the verb 'rain' has an implicit location argument – it never just rains, but it rains somewhere. By contrast, the verb phrase '=4', expressing the property *being identical to the number four* has no such implicit arguments: mathematical truths are not location sensitive in this way. That is to say, the presence or absence of these adjuncts is determined by the demands of semantics, and so it is undermotivated from the point of view of syntax.

So here is where we stand. We first considered the argument for positing covert temporal variables. We then went on to consider another case of positing covert (domain restricting) variables, and a criterion for when it is licit. After that, we saw reasons to doubt the syntactic legitimacy of such variables. Let us grant that Collins' argument against Stanley holds, and, as

---

[3] As he notes, this is a different to positing items like PRO or traces, which have stronger syntactic grounding.

he himself surely thinks, generalises beyond that one case. Collins therefore thinks, naturally enough, that if these theorists are positing syntax that is not there, then their theories are falsified. It is this that I want to challenge in the remainder of the paper. Even granting that Collins is right about syntax (which I independently do), his conclusions about semantics can be resisted. And in particular, turning to the tense debate, which Collins was not discussing, let us assume that positing of temporal variables (for example) is as syntactically unmotivated as positing domain restricting variables. If that is so, the Collins line would have it that because positing temporal variables is no good, so, by extension, the quantifier view of tense is also no good. The remainder of the paper attempts to argue against this, showing that one can reconcile the semantic criterion for positing variables with the demands of syntax.

4. Why there is no cause for concern about the above

For the remainder of the paper, I want to try to dispel any worries caused by the tension I have just created. The first step of my argument goes as so: if positing covert stuff is misguided, semantics is misguided. The second step argues that semantics is not misguided. Accordingly, positing covert stuff is not misguided, and so quantifier theorists of tense need not worry that by positing covert temporal variables they are doing something wrong.

4.1. If positing covert stuff is misguided, semantics is misguided ...

My basic worry with anti-positing covert stuff arguments is that they threaten to impugn an awful lot of work on semantics; indeed, followed through, if we take a hard line against such positing, we are left with next to none of the foundations of the discipline that has arisen in the last 50 or so years.

My argument is quite simple. Its central premise is that objectual quantification, as typically understood, involves positing covert elements. The most easy way to see this is to note its dependence on *indexes* conceived of as bits of syntax to achieve its aims, although I think the point applies also to the lambda terms used for binding, as well, although perhaps less forcefully, to the traces posited as arguments to verb phrases.

However, that is to get ahead of myself. I better, at least briefly, review the textbook theory. Consider a sentence like the following (treating

'Everybody' as a single word so we don't have to bother with a story about its composition):

(15) Everybody is happy.

Our goal is to give a compositional categorematic theory, by which I mean one which assigns to each expression in the sentence a meaning and determines the meaning of sentences and other complexes of words on the basis of their parts using as few rules as possible. I assume we have already seen the rationale for treating 'is happy' as a function from objects to truth values, notated $<e, t>$, and for assuming that it combines with – for example – the denotations of name arguments by means of function application.

The neatest thing would be to say the same here: that the parts are composed by function application. Given that, and assuming 'Everybody' is not type e, the only option is to treat it as a function from functions to objects to truth values to truth values, that is, notated, of type $<<e,t>,t>$.

However, things must be more complicated. Consider:

(16) Datri loves everybody

Assuming 'loves' is $<e,<e,t>>$, then we will get a type mismatch trying to compose the verb phrase. 'Loves' looks for an e but finds an $<<e,t>,t>$. So what we do is, we assume the possibility of movement. An expression can appear at LF in a place other than where it appears on the surface, and when it moves, it leaves behind what is called a trace. This gives us:

- Everybody Datri loves t_1.

That does not help overmuch. If 't_1' is type e, then the lower sentence 'Datri loves t_1' should denote a truth value, rather than the desired $<e,t>$ (which is what the quantifier looks for). So we use another device, a device of abstraction which turns a sentence into a function by treating the trace as a variable and having this device bind it. We then get:

- Everybody λt_1 Datri loves t_1

Now we have the right typing, and composition can proceed as we would like it to. However, think of how far we have deviated from the surface syntax of the sentence – movement, lambda binders and indexes are now posited as part of the syntax where they were not before.

Now, the motivation for movement is arguably solid, or, if it is been rendered troublesome by recent developments in syntax, is at least an instance of using a syntactically responsible methodology (since it seems independently required to make sense of such mundane sentences 'Who did you meet?' where the object of 'meet', 'who' appears to have moved from its position in the argument place of the verb to the front of the sentence). The other items are less motivated. Indeed, in a more recent paper (Collins 2017), Collins explicitly makes this point with regard to lambda binders: they are 'parachuted into the syntax with no independent rationale' (13) merely to get the typing for quantification right. The same thing applies to the indexes on variable-like elements.

I think this last issue, especially, can often be overlooked because of its familiarity, but itis worth emphasising just because it shows that positing comes in very early to our semantic education. In the Heim and Kratzer treatment, it is not only the invisible traces that have indexes; so do pronouns. But this is to make quite a leap from what the surface syntax gives us. Pauline Jacobsen makes the point forcefully:

> An ... important observation – and one which I feel is worth heavily stressing – the fact that pronouns are not literally variables. After all, we don't say 'He-sub-i saw him sub-j'. Rather (modulo number, gender and case), we have an invariant pronoun which – quite unlike a variable – does not come in the phonology with an index. Crucially, variables are distinct from each other, but pronouns are not. Thus the common wisdom that pronouns transparently reveal themselves as variables is simply incorrect. (Jacobson 1999, 145, emphasis in the original)

I agree this is worth heavily stressing.[4] Pronouns are not variables, yet to account for even simple cases of pronominal binding, according to the textbook we use to introduce semantics to our students, we need to assume that they are, and in particular that they carry unpronounced numerical indices. And pronominal binding is surely an essential part of the theory of natural language quantification, and the theory of natural language quantification is a fundamental, deeply rooted part of semantic theory. Positing covert items, then, goes deep: it is not just something we need to account for recherché sentences involving tenses (not that such

---

[4] A referee points out that indexes have come under attack in minimalist frameworks in the last few years, so if youare of a particularly minimalist bent and none of this moves you, just consider the same argument as in the text targeting the lambda binder.

sentences are typically recherché: tense is as fundamental a feature of language as quantification).[5]

What is the point of the discussion in this section? It is that positing covert syntax to get one out of a bind (pun intended) is pretty crucial to prevailing semantics methodology. If we were forced to give it up, it would eat into the very heart of our textbook semantics: indeed, pretty much anything past Heim and Kratzer chapter two would be out, which is to say most of semantics would be out.

4.2. ... But semantics is not misguided

That would be very bad news. In fact, though, I think the magnitude of its badness comes to our rescue. It cannot be, I think, that most of semantics is out. Granted that if positing covert variables is misguided, semantics is misguided is true, the thing to do is tollens on that conditional.

That is not just cockeyed optimism. Formal semantics seems like it is doing quite well: in the past 50 or so years, our understanding of a wide range of semantic phenomena has increased dramatically. We have developed sophisticated technical apparatuses to deal with increasingly larger fragments of natural language, and devised new theories using them to account for ever larger amounts of linguistic data. Moreover, these theories are falsifiable, and indeed frequently falsified. And they are constrained, not only by the data, but by theory-internal criteria. We look for theories that are technically simple, and this applies also to posited things: too many posits, and readers will tend to balk at a theory. It is not the case that anything goes.

I think we should take this productivity as a – defeasible, surely – sign that semantics is not fundamentally broken. And to the extent that many of these advances depend on a theory which posits covert items, we should view these advances as an indirect justification of that practice.

That is pretty much all there is to my argument. I will respond to some perhaps slightly less obvious objections in the next section but for now I want to consider a more salient one. I say that positing variables is motivated, but syntax, as practiced by syntacticians, says that it is not. Surely the syntactician should win – how do I square my position with the disagreements of actual researchers in the field?

---

[5]Again, I should say it goes deep *according to the most standard textbooks we teach from and the assumptions we tend to start from in doing semantics*. I certainly do not mean to say that one is compelled to treat pronouns like variables; indeed, in a couple of pages we will see an alternative approach that does not require this. Just that, as a matter of sociological fact, most working semanticists and philosophers of language tend to treat them as such.

I do so by saying that, despite what we say and write when we write down trees and posit items, we are not *really* doing syntax. We are doing something else. There is some feature of linguistic reality related to meaning and in some sense, the structures which we posit describe it. This naturally suggests two questions: what is this bit of reality? Granted it is not syntax, why think it exists?

As to the first question, I suggest that we do not know what it is. Some of us might think that it is syntactic reality, but we are wrong to do so, as the syntacticians tell us. But that does not matter – it is not necessary, to study a bit of reality, that one have a good conception of what it actually is. Our lives both scientific and otherwise are replete with dealings with things the real natures of which are opaque to us.

Thus, to repeat my overblown example from earlier, we really have no idea what is going on when we use the Schroedinger equation to describe the evolution of a physical system, but we still do use it, because it works. People working at the foundations of physics can and do ask what is *really* going on, and in so doing produce different interpretations of the formalism, but for most people it does not really matter. It is not really an objection to a physicist using the equation to say that the quantum world is completely mysterious and they do not know what they are talking about. Similarly, it is not an objection to the semanticist to say that positing covert variables is mysterious. It seems to get the job done, and that is enough.

I do not mean this to be a sort of instrumentalist or anti-realistic position. There really is some bit of reality out there that is described, somehow, by the Schroedinger equation. We just have no idea what it is like. Similarly, in drawing our trees and positing our variables, we are latching on to a piece of linguistic reality; we just do not know what it is really like (the one thing we know, perhaps, is that it is not fundamental syntax). And, again, we have the assurance we are latching on to something by the effectiveness of semantics.[6]

5. Some objections and replies

In this section, I want to consider briefly some more objections to the argument presented above.

The first objection pertains to the difference between positing object language variables and positing temporal variables.[7] My argument

---

[6] Let me just make it clear: I am under no illusions that the effectiveness of semantics is anywhere near that of quantum physics.
[7] I thank an anonymous reviewer for this objection.

basically went as so: if you think positing temporal variables is bad, you should think positing object denoting variables is bad too. But you should not think the latter so you should not think the former. But the conditional can be questioned. Is it not a perfectly coherent position to think that positing object variables is okay, but positing temporal variables is not, because there is more evidence for the object variables? As already noted, there are items that appear overtly in English and other languages that behave like variables in some respects, namely pronouns, and there is some syntactic reason for positing traces (namely movement). Given that, you might not think that adding lambda binders and indexes is not such a big deal. By contrast, there does not appear to be any such evidence for temporal variables: they seem to be appearing from nowhere, and there seems to be no antecedent reason to believe that they exist. So one might think the two cases are just different, and so we cannot go from the claim that there are no temporal variables to the claim that there are no object variables.

I take this point, and think it is a reasonable objection to the view that I have put forth here. I think one's attitude towards this matter will depend on how strongly one weighs the respective evidence: if you are impressed with the claim that there are object variables, then the further posits of Heim and Kratzer might not seem so bad. On the other hand, if you are not impressed with that claim, for example because you think pronouns are notably different from the variables of logic, or because, say, movement data do not, well, move you, then you will object to lambda binders and indexes. I belong to the latter camp, and so I think the syntactic bona fides of object and temporal variables is about the same, but I do not think the former view is absurd.

Let me turn to another objection. I argued that positing variables and covert structure in general was acceptable because it was needed to get semantics off the ground. But, of course, some argue that it is *not* needed for this purpose. Indeed, I quoted with approval perhaps the most famous such semanticist, Polly Jacobsen, whose 'variable-free' semantics is animated, at least in part, by the passage quoted above, as well as by a distrust of movement and other such covert syntactic operations. She seeks to provide a direct-compositional semantic theory, where that is one that runs off something like the surface structure of the sentence.

Accordingly, one might object that the lesson to be learned from all this is simply that we should opt for her semantics. However, I do not think that someone animated by fears about semantic's overreach into syntax ought to take much comfort from Jacobsen's theory, because, at least as she typically presents it, it involves a lot of semantically guided syntactic revisionism as well.

To make this point, it is necessary to say something briefly about the very basics of her system. For the standard theory, as we have seen, pronouns denote objects relative to assignment functions; binding is then a question of abstracting away their assignment sensitivity. For Jacobsen, on the other hand, pronouns denote identity functions which then combine by means of function composition. In function composition, one takes a function F from type a to type b and G from type b to type c and forms a new one from type a to type c defined as so $\lambda s_{<a>}.G(F(s))$.

Thus consider a very simple sentence:

(17) Mary loves him

Composing the verb phrase, we will have

- [$\lambda y. \lambda x.x$ loves y] ° [$\lambda x.x$] = [$\lambda y. \lambda x.x$ loves [$\lambda x.x$](y)]

However, there is an immediate concern here. The problem is that if we were to do function application in the next step, we would end up with Mary as the lovee rather than the lover. But function composition will not work either because 'Mary' denotes an object of type e, rather than a function.

If we were to type-raise Mary to its generalised quantifier denotation, so that it denoted $\lambda P.P(Mary)$, and then if were to use function composition on the $<e, <e,t>>$ with that $<<e,t>,t>$ yielding (the not very fun to read) $\lambda y.\lambda P(Mary)[\lambda x.x$ lost [λx](y)] we will get the right results (provided we assume the sentence can denote a function whose input is given by context, which seems like a plausible way to treat the deicticly functioning pronoun 'him').

However, the question is how does the semantics *know* that this type-raising is the right thing to do? And Jacobsen's answer is that it is in the syntax. She posits a syntactic type-raising operation so that strictly speaking, 'Mary' as it occurs in that sentence does not belong to the syntactic type of determiner phrases but rather a special syntactic type for generalised quantifiers. In general, to get her semantics playing nice with her syntax, Jacobsen posits a wide range of syntactic categories such that, for example, pronouns and names, strictly speaking, do not belong to the same category.

I will not go into all the details: the point, finally, is that Jacobsen's theory is just as syntactically revisionary as Heim and Kratzer's: it just makes different revisions. So we cannot get round the problem of positing covert stuff by switching to avariable-free theory.

Let me turn to a third objection, which is a more general worry about the nature of my argument. My argument is based on a very sanguine conception of formal semantics, namely that it is a productive research programme that is making progress. You might wonder what my evidence for this claim is. It is not like the example I used earlier, of physics, in which we have a guarantee that we are on to something because, for example, we can build machines that work because they make use of the theory. Semantics is less testable in this way. Moreover, you might think that there have been plenty of research programmes the proponents of which thought were making progress but was in fact not. Smart people sat around learned societies and cafes, just as we sit around universities, developing their theories about phrenology or Lacanian psychoanalysis and arguing, perhaps ingeniously, with each other. But they were completely misguided in their belief that they were making progress in accounting for personality or neurosis. Who is to say we are any better?

Ultimately, I do not think I can conclusively argue against this here. It is a possibility. What I can do is reiterate what I said before: the theories we develop attempt to take into account a range of often cross-linguistic data, while striving for simplicity and technical competence, and being subject to criticism by many smart coresearchers. It certainly *seems* as if these are exactly the sort of virtues one would want a theory of a bit of reality to exhibit, and so I would be tempted to say that the onus is on the opponent of formal semantics to make their case.

The fourth objection, which I owe to a reviewer for this journal, complains that the view developed here rides roughshod over important distinctions between semantics and pragmatics. Here is one way to put it: the view that I have been defending seems to be obliquely committed to the claim that wherever we have the semantic effects of binding, we have its standard syntactic realisation.

That is an uncomfortable commitment to have. It compels one, to take but one example, to say that donkey and discourse anaphora are to be accounted for syntactically. But there are, as is familiar, a range of viable theories of such anaphora that make no claims about syntax but which rely, on some level, on pragmatics (examples include Neale 1990; Ludlow and Neale 1991). More generally, it is just false that when a sentence exhibits some semantic behaviour, that behaviour must be accounted for syntactically. Pragmatic options are available. A theory that does not respect that fact is no good.

I agree. I am not defending a general semantic behaviour implies syntactic realisation view. Pragmatic options are indeed available. I am committed to the narrower point about binding, and in that case, I *am* defending the claim that semantic behaviour implies syntactic realisation. That leaves me with a not unsubstantial dangling commitment: to defend a syntactically realised semantics of donkey and discourse anaphora. But thankfully the literature can come to my aid, and I can point to Elbourne (2005) as an example of such a theory.

Finally, the fifth objection, also owed to a reviewer for this journal, has it that by turning my back on syntax as it is practiced by working syntacticians, I have thereby deprived myself of a useful if partial means of developing and deciding between competing theories. Surely, the reviewer says, it is not the case that *anything* goes when we are writing down our trees in our papers. But if we are not required to ensure that our trees are compatible with what contemporary syntax tells us, what constraints *are* we obliged to follow?

Well, that is a very good question, and one to which I do not have a full answer. But I think there are some things we can say, things which in fact reflect our perhaps unreflective current practice. Generally speaking, less is better. We are inclined, I think, to judge theories more harshly the more they rely on covert posits. If someone's theory of whatever semantic phenomenon requires wildly complicated trees with tons of hidden type-shifters, we are inclined to disbelieve that theory. Generally speaking again, neutrality is useful. If a given semantic theory can only be cashed out, say, in Pietroski's (2005) syntactic-semantic architecture, then we should be less inclined to believe it than one that can also be implemented using Heim and Kratzer's assumptions, or again Jacobsen's. More generally, there are criteria of simplicity, explanatory power, empirical adequacy and so on that we use when deciding theories and I think we can appeal to them when drawing our trees, so that it is not the case that anything will go.

6. Conclusion

The quantifier theory of tense depends crucially on positing covert items in syntax. This paper has considered an obvious objection to it based on this fact: that in making such posits the semanticist is overreaching and doing, poorly, the job of the syntactician, and we have no independent syntactic grounds for thinking that such items exist. My response, in essence, is that positing covert items is too big to fail: too much of semantics rests on it for it to be misguided. This is not, I suggested, merely groundless optimism. We should be confident in the methodology of positing because semantics seems to make progress and that methodology is central to semantics. Moreover, I argued for a sort of agnosticism about it is we are doing when we posit covert items – granted that it is not fundamental syntax, what is it? I suggested that we do not know, but that this is not an objection, because we frequently study things the nature of which we do not understand. The quantifier theorist, then, need not worry about the syntactic commitments their theory makes.

Acknowledgements

The author would like to thank two anonymous referees for *Inquiry* for several helpful comments and for clearing various points up. In addition, the author would like to thank the participants at the Operator vs. Quantifiers conference in Barcelona this special issue is based on; during its three days, I learned more about the debate than in years of reading. The author also thanks Max KIbel and David Rey, both for organzing the conference and for editing this special issue. Finally, thanks to Matthew Cameron, for many hours of fun discussions about this and related topics over various years in various countries.

Disclosure statement

No potential conflict of interest was reported by the author.

References

Cappelen Herman, and Ernie Lepore. 2002. "Indexicality, Binding, Anaphora and a Priori Truth." *Analysis* 62 (4): 271–281.
Collins John. 2007. "Syntax, More or Less." *Mind* 116 (464): 805–850.
Collins John. 2017. "The Semantics and Ontology of the Average American." *Journal of Semantics* 34 (3): 373–405.
Cresswell, M. J. 1991. *Entities and Indicies*. Dortrecht: Kluwer Academic Publishers.
Dowty David R. 1982. "Tenses, Time Adverbs, and Compositional Semantic Theory." *Linguistics and Philosophy* 5 (1): 23–55.
Elbourne, Paul D. 2005. *Situations and Individuals*. Vol. 90. Cambridge: MIT Press.
Fara, Delia Graff. 2015. "Names are Predicates." *Philosophical Review* 124 (1): 59–117.
Heim, Irene, and Angelika Kratzer. 1998. *Semantics in Generative Grammar*. Vol. 13. Oxford: Blackwell.
Jacobson Pauline. 1999. "Towards a Variable-free Semantics." *Linguistics and Philosophy* 22 (2): 117–185.

Kamp Hans. 1971. "Formal Properties of 'Now'." *Theoria* 37 (3): 227–273.

King Jeffrey C. 2003. "Tense, Modality, and Semantic Values." *Philosophical Perspectives* 17 (1): 195–246.

Ludlow Peter, and Stephen Neale. 1991. "Indefinite Descriptions: In Defense of Russell." *Linguistics and Philosophy* 14 (2): 171–202.

Moss, Sarah. 2015. "On the Semantics and Pragmatics of Epistemic Vocabulary." *Semantics and Pragmatics* 8 (5): 1–81.

Neale, Stephen. 1990. *Descriptions*. Cambridge: MIT Press.

Partee Barbara Hall. 1973. "Some Structural Analogies between Tenses and Pronouns in English." *Journal of Philosophy* 70 (18): 601–609.

Pietroski, Paul M. 2005. *Events and Semantic Architecture*. Oxford: Oxford University Press.

Prior, A. N. 1968. *Papers on Time and Tense*. Oxford: Oxford University Press.

Schlenker Philippe. 2006. "Ontological Symmetry in Language: A Brief Manifesto." *Mind and Language* 21 (4): 504–539.

Stanley Jason, and Zoltán Gendler Szabó. 2000. "On Quantifier Domain Restriction." *Mind and Language* 15 (2&3): 219–261.

Vlach Frank. 1973. "'Now' and 'Then': A Formal Study in the Logic of Tense Anaphora." PhD thesis, University of California Los Angeles.

Looking backwards in type logic

Jan Köpping and Thomas Ede Zimmermann

ABSTRACT
Backwards-looking operators Saarinen, E. [1979. "Backwards-Looking Operators in Tense Logic and in Natural Language." In *Essays on Mathematical and Philosophical Logic*, edited by J. Hintikka, I. Niiniluoto, and E. Saarinen, 341–367. Dordrecht: Reidel] that have the material in their scope depend on higher intensional operators, are known to increase the expressivity of some intensional languages and have thus played a central role in debates about approaches to intensionality in terms of implicit parameters (as in modal and tense logic) vs. variables explicitly quantifying over them. The current contribution takes a look at these operators from a type-logical perspective. It is shown that extending Gallin's ([Gallin, D. [1975. *Intensional and Higher-Order Modal Logic*. Amsterdam: North-Holland Pub. Company]) translation from intensional type logic (IL, Montague, R. [1970. "Universal Grammar." *Theoria* 36: 373–398]) to two-sorted type theory so as to include a version of Yanovich's (Yanovich, I. [2015. "Expressive Power of *Now* and *Then* Operators." *Journal of Logic, Language and Information* 24: 65–93]) backwards-looking operators, does *not* increase the expressive power of formulae with exclusively intensional parameters. The result, which makes use of a theorem by Zimmermann (Zimmermann, T. E. [1989. "Intensional Logic and Two-Sorted Type Theory." *The Journal of Symbolic Logic* 54 (1): 65–77]), is illustrated by pertinent examples from the literature. The paper closes by indicating alternative strategies of incorporating backwards-looking operators.

Come, let Us go down and there confuse their language, that they may not understand one another's speech

(Gen. 11:7)

1. Setting the stage

Whenever co-referring terms, co-extensional predicates or materially equivalent sentences cannot be substituted *salva veritate*, they appear in *intensional* environments. There are two approaches to the representation of intensionality in formal logic. On the first approach, favored by modal and tense logic, the denotations of all formulae depend on certain parameters, and intensional environments are marked by special operators that shift the denotational dependence of anything in their scope. On the second approach, which has become more and more popular among semanticists, the same parameters are represented by special variables that are bound by quantifiers (or, more generally: λ-abstracted) to create intensional environments. Since the variable binding approach merely explicitly names the parameters the operator approach relies on, the choice between the two may seem a matter of notational convenience. However, the expressive means of explicit variable-approaches usually cover constellations beyond those obtained by spelling out implicit parameters in operator-based formalizations. In fact, the exact relationship between the two approaches is somewhat tricky. Before going into the purported gain of expressivity offered by backwards-looking operators, it is thus worthwhile to take a quick look at that relationship in general. As it turns out, the size and nature of the expressivity gap between implicit parameters and explicit variable binding cannot be assessed by comparing individual formulae in isolation but needs to take their environment into account – and more specifically it needs to consider:

(a) the formal languages to which the formulae compared belong;
(b) the choice of the determinants of denotation;
(c) the relevant notions of denotation to be preserved in that comparison.

These general observations will be of immediate importance both to the general comparison of backwards-looking operators in intensional and two-sorted type logic and to the discussion of concrete examples in Section 2. A few remarks on (a)–(c) are thus in order.

Concerning (a), it should first be noted that a system of implicit indices does not by itself determine a unique system of explicit variable binding. Modal propositional logic is a case in point. It is common and natural to construe (2) as an explicit counterpart of (1), which is why, following Fine (1975, 19), (2) is commonly called the *standard translation* of (1):

(1) $\Box \Diamond p \to \Diamond \Box p$
(2) $[(\forall i_1)] i_0 R i_1 \to (\exists i_2)[i_1 R i_2 \wedge P(i_2)]] \to$
    $(\exists i_1)[i_0 R i_1 \wedge (\forall i_2)[i_1 R i_2 \to P(i_2)]]$

In (2), the variables i_n range over worlds (or *indices*), with i_0 indicating the value of the evaluation parameter; the unary predicate P represents the propositional variable p of (1); and R stands for the accessibility relation that is implicit in a standard interpretation of (1). Hence (2) straightforwardly spells out the truth conditions of (1) in a (pointed) Kripke model; it is not hard to see (and well-known) that predicate logic versions along the lines of (2) can be given for all formulae of modal propositional logic.[1] So the pertinent expressivity question seems to concern modal propositional logic vs. first-order predicate logic with one binary and infinitely many unary predicates – or M_0 vs. L_1, for short. And the two languages differ in expressivity: not every L_1-formula is equivalent to the standard translation of a formula in M_0 – a case in point being (3), the M_0-inexpressibility of which can be shown by basic model-theoretic techniques:[2]

(3) $(\forall i_0)\, i_0 R i_0$

On the other hand, being a first-order formula, (2) is *a fortiori* a formula of second-order logic (with first-order predicates), which allows for a whole lot of additional distinctions beyond L_1. In that sense, the size of the expressivity gap between modal formulae and their explicit counterparts trivially depends on the 'target' language that the latter are supposed to represent.

In a similar vein, the 'source' language is not determined by one given modal formula either. Higher-order modal logics with universal (or unrestricted) accessibility are particularly revealing in this respect: what is expressed by a quantification over propositions as in (4) may depend on whether (i) the interpretation of the quantifier is restricted to standard models, or (ii) generalized (Henkin) models are admitted: in case (i), (4) boils down to a cardinality condition on Logical Space, W, viz. that it contains precisely one world; (ii), on the other hand may (but need not) be set up so that W is always denumerable.[3] In any case, the standard translation (5) of (4) does express said cardinality condition whether or not it is construed in a standard way.

(4) $(\forall p)[\Box p \vee \Box \neg p]$
(5) $(\forall P)[(\forall i_0)P(i_0) \vee (\forall i_0)\neg P(i_0)]$

(b) ought to be familiar from modal propositional logic too. For, apart from being first-order translatable as (2), (1) may also be seen as *corresponding to* the following formula of monadic second-order logic:

---

[1] More precisely, (1) is true in (W,R,V,w) just in case (2) is true in a corresponding first-order model (W,F) under a corresponding assignment h, where $F(R) = R$, $F(P) = V(p)$, and $h(i_0) = w$.
[2] More specifically, *unfolding* makes the accessibility relation of a Kripke model irreflexive, but preserves the truth values of modal formulae; cf. Goranko and Otto (2007, 262).
[3] See Gallin (1975, 30ff.) for a pertinent construction in intensional type logic.

(6) $(\forall P)[(\forall i_1)[i_0Ri_1 \rightarrow (\exists i_2)[i_1Ri_2 \wedge P(i_2)]] \rightarrow$
     $(\exists i_1)[i_0Ri_1 \wedge (\forall i_2)[i_1Ri_2 \rightarrow P(i_2)]]]$

According to this construal, the propositional letters of M_0 correspond to predicate *variables* (as in the standard translation (1) of a higher-order formula). The crucial difference between the mere standard translation (2) and the corresponding formula (6) is that the former takes (1) to (partially) describe a particular world in a given model, whereas the latter construes it as (partially) describing Logical Space: (2) is correct in that it reflects the truth value of (1) across all pointed models (W,R,V,w), whereas (6) reflects its behavior across all frames (W,R). The two standards of correctness thus differ in what counts as determining the truth or falsity of the modal formula: the standard translation takes a modal formula to be true or false given a pointed Kripke model (W,R,w), whereas correspondence construes it as being about accessibility in Logical Space:

(7) φ is *true in* a model (W,R,V) at a world $w \in W$ iff $[\![\varphi]\!]^{(W,R,V,w)} = 1$.
(8) φ is *true throughout* a frame (W,R) iff φ is true in all models (W,R,V,w).

In particular, the valuation V and the choice of the actual world w count when it comes to interpret the modal formula in the spirit of standard translation, but they are irrelevant for modal correspondence. Of course, both construals of modal formulae are perfectly legitimate and neither is in any sense superior to the other,[4] but it is important to distinguish between them when it comes to exploring and comparing expressivity. As a case in point, (1), which is known as *McKinsey's axiom*, is not even first-order expressible if construed like (6); cf. van Benthem (2001, 351f.).

In the next sections, we will only be concerned with local construals of type-logical formulae. However, we will make use of Kaplan's (1979) two-dimensional treatment of context dependence, according to which the truth value $[\![\varphi]\!]^{M,c,i}$ a model M assigns to a formula φ depends on a context c and an index i.[5] As a consequence, two kinds of truth conditions emerge (within M):

(9) φ is *true at* a point of reference (c,i) iff $[\![\varphi]\!]^{M,c,i} = 1$.
(10) φ is *true in* a context c iff $[\![\varphi]\!]^{M,c,i_c} = 1$.

---

[4] Apart from these local and global construals, there is also a 'regional' interpretation of modal formulae that generalizes over worlds but not models; cf. van Benthem (2001, 327f.).
[5] This includes the special case that the sets of contexts and indices both coincide with the set of possible worlds. In this case, the context may be construed as the world-component of a pointed Kripke model.

In (10), i_c is a unique index suitably determined by the context c; in the (formally) simplest cases i_c may be identified with c so that the unary truth defined in (10) boils down to truth on the diagonal of all pairs (c,c). Now, in order to compare such a two-dimensional system with a given system of explicit indices, one first needs to decide which of the truth conditions the latter ought to capture: if (10) is all that counts, the context would be the only parameter to be represented by a (globally) free variable; if however, (9) is at stake, then context and index would have to be represented by two separate variables. On the one hand, a more ambitious and fine-grained account might aim at capturing truth *at* a point of reference; on the other hand, truth *in* a context might be all that is pre-theoretically available.[6] As far as we can see, both options make sense, but they need to be distinguished. As in the case of standard interpretation vs. correspondence, the difference bears heavily on matters of expressivity. As a case in point, the two-dimensional *actuality* operator defined in (11) does not increase the expressive power of M_0 if measured by (10),[7] though it trivially increases expressivity in terms of truth as defined in (9):

(11) $[\![\mathbf{A}\varphi]\!]^{M,w,w'} = 1$ iff $[\![\varphi]\!]^{M,w,w} = 1$

To see the impact of the difference, one may compare any formula φ with its actuality variant $\mathbf{A}\varphi$, which are true in the same contexts (in any model) but may come apart at other points of reference.

The double indexing characteristic of two-dimensional semantics must not be confused with the splitting of indices into components. In modal logic, indices coincide with possible worlds, in tense logic they are times, and if the two are combined, indices come out as world-time pairs, thus allowing for reference shifts to past or future situations, both counterfactual and actual. To make such structured indices explicit, one may have variables either for each of their components or for entire tuples. The choice between the two options depends on the index-shifting operators but, as far as we can see, does not bear on expressivity: if a tense operator only affects the time-component of the index (as in the interpretive clause for the $\mathbf{P}_{[AST]}$ operator in (12)), its most natural explicit

---

[6] In other words, one may question the theory-external significance of propositions expressed by sentences. See Lewis (1980, 96f.) for some (brief) remarks and Dummett (1981, 565f.) for a (lengthy) discussion of a related point pertaining to Kripke (1972). See also Rabern (2012) and Zimmermann (2012, 2403f.) on this perspective.

[7] Hazen, Rin, and Wehmeier (2013); see also Hodes (1984) and Wehmeier (2003) for the corresponding negative result on modal predicate logic.

translation involves a quantifier that binds a time variable (as in (13a)); but then again, it may also be represented by a quantifier over complex indices $i = (i_1, i_2, \ldots)$ (as in (13b)):[8]

(12) $[\![\mathbf{P}\varphi]\!]^{(w,t,\ldots),g} = 1$ iff $[\![\varphi]\!]^{(w,t',\ldots),g}$, for some $t' < t$
(13) a. $(\exists t_1)[t_0 S t_1 \wedge \varphi[t_0/t_1]]$ where $[\![xSy]\!]^g = 1$ iff $g(y) < g(x)$
 b. $(\exists i_1)[i_0 R i_1 \wedge \varphi[i_0/i_1]]$ where $[\![xRy]\!]^g = 1$ iff $g(x)_1 = g(y)_1$ and $g(y)_2 < g(x)_2$

In the examples below, we will follow the formalization strategy in (13b) and also extend it to contexts, which may be split up in an analogous (though not necessarily identical) fashion.[9] We regard the difference between the two explicit versions (13) of (12) as largely one of mere notational convenience, whereas the choice between representing points of reference vs. contexts is substantial.

Finally turning to (c), we note that so far we have only looked at truth-evaluable formulae. Of course, this suffices as long as the only recursive categories are sentential, as in propositional and predicate logic. However, when it comes to the analysis of natural language meaning, richer systems of functional type logic are employed that allow a stepwise, *compositional* formalization of sub-sentential expressions. As a case in point, while the sentence (14) can be expressed in first-order predicate logic as in (15a), its subject cannot, and neither can the quantificational determiner *every*. However, once typed terms defined by lambda-abstraction are available, compositional translation can proceed by translating the quantifier and the determiner as (15b) and (15c), respectively:

(14) Every horse neighs.
(15) a. $(\forall x)[H(x) \to N(x)]$
 b. $\lambda P.(\forall x)[H(x) \to P(x)]$
 c. $\lambda Q.\lambda P.(\forall x)[Q(x) \to P(x)]$

All three formulae in (15) are terms of Montague's (1970) intensional type logic *IL*, a version of which will be introduced in Section 2.2. At this point it suffices to realize that only (15a) denotes a truth value; the denotation of (15b) is the set of all sets of individuals that contain the extension of

---

[8] In (13b), i_n is the nth component of index i and $\varphi[x/y]$ is the result of replacing free x in φ by y. Note that the domains of the assignments in the two cases in (13) are different.
[9] For the relation between context- and index-splitting see Zimmermann (2012, 2399ff.), where both are subsumed under the term *parameterization*.

H[orse] as a subset, whereas (15c) stands for the subset relation between sets of individuals.[10] This difference in denotation is reflected in the (syntactic) *types* of the three terms, which are, respectively: t; $\langle\langle e,t\rangle,t\rangle$; and $\langle\langle e,t\rangle,\langle\langle e,t\rangle,t\rangle\rangle$.

IL is an intensional logic interpreted by implicit world and time parameters, and its most obvious explicit counterpart is the language *Ty2* of *two-sorted extensional type theory*,[11] which has even more types of terms most of which are arguably dispensable for purposes of linguistic semantics. Still, the natural question arises as to whether the expressivity of the two languages should be compared only on the truth-valuable terms of type t, or whether all types common to *IL* and *Ty2* should be taken into consideration. As in the decisions concerning (a) and (b) above, there is no right or wrong here, but the matter should be clear from the start; for again it does bear on expressivity. In fact, Gallin (1975, 105) has shown that the closed *IL*-terms of type t cover whatever is expressible in its explicit counterpart *Ty2*.[12]

However, Gallin's result has no bearing on open formulae and, more importantly, terms of types other than t. In fact, it is obvious that it does not extend to the full range of *Ty2*-terms, which are based on a richer type hierarchy than *IL* is: *Ty2*-terms of non-intensional types cannot be equi-denotational to any *IL*-terms; nor can those *Ty2*-terms that contain parameters (= free variables or constants) of such types. One may think that differences in denotational expressivity are irrelevant as long as the two languages are truth-conditionally equivalent. Concerning the relative denotational expressivity of *IL* and its explicit variable version *Ty2*, closer inspection reveals that the superiority of the latter is largely due to differences in the type-hierarchy and appears to be of limited relevance to compositional interpretation. In particular, it does not show in formulae of intensional types and with intensional parameters (cf. Zimmermann 1989). The reason is that the lack of index-variables can frequently be made up for by resorting to higher-order abstraction. As a case in point, the *Ty2*-formula (16a), which denotes (a Curried version of) identity

---

[10] We follow the common practice of identifying sets with their characteristic functions whenever this is unlikely to lead to confusion.

[11] Actually, *three-sorted* type theory, which takes worlds and times to constitute separate sorts instead of combining them into indices, would do just as well. However, the more common *Ty2* allows us to directly apply known results to *IL*-extensions investigated below.

[12] Though Gallin did introduce *Ty2* as the natural variable-binding counterpart of *IL* (and also gave a translation algorithm from the latter to the former), it is clear that Montague's (1970) very construction of *IL* involved, as it were, cutting *Ty2* to size. So, contrary to what is sometimes suggested in the literature, Gallin can hardly be seen as having discovered the relationship between the two systems, let alone invented *Ty2* – neither did he claim to have done so.

between indices can be rendered in *IL* by the more cumbersome but equivalent (16b):[13]

(16) a. $[\lambda i_0.[\lambda i_1.[i_0 = i_1]]]$
    b. $[\char94[\lambda F.[F = \lambda p.\char94 p]](\lambda p.\char94 p)]$

Thus, obvious differences notwithstanding, both truth-conditionally and denotationally, *IL* turns out to be surprisingly close in expressive power to its explicit counterpart *Ty2*. In the next section, we will show that backwards-looking operators do not disturb this general picture. To this end, we will adapt and generalize Yanovich's (2015) model-theoretic characterization of these operators to a two-dimensional version of Montague's (1970) intensional type logic and apply a variant of Gallin's (1975) standard translation to it. As a result, the 'quasi-closed' formulae of type logic with backwards-looking operators turn out to be in a fragment of *Ty2* which is known to be expressible in the original language *IL*. Whatever the impact of this result, it does not mean that there is a 'pointwise', compositional way of defining away backwards-looking operators in terms of intensional type logic.

2. Backwards-looking operators in intensional type logic

2.1. Examples

The potential increase of expressive power that comes with the introduction of backwards-looking operators $\boldsymbol{\partial}_i$ can be illustrated with examples like (17), which is part of a straightforward extension of M_0:

(17) $\Diamond\Diamond\, (\varphi \wedge \boldsymbol{\partial}_1 \psi)$

The idea behind these operators is that they exempt their argument from being evaluated at the current (possibly locally bound) index and instead have it evaluated at a higher index, either the outermost, free index of evaluation or at any other one that is introduced and bound by an operator in a structurally higher position. Thus, the outer $\Diamond$ in (17) shifts the point of evaluation from some w_i to the w_i-accessible worlds w_j, from which certain worlds w_k are accessed to evaluate the embedded conjunction. However, whereas the conjunct φ is evaluated at w_k, ψ is evaluated at w_j bound by the outermost $\Diamond$, because it is in the scope of the backwards-

---

[13]See Section 2.2 for notational details.

looking $\mathbf{a}_1$: in (17), $\mathbf{a}_1$, by way of its subscript, refers back to the world bound by the outermost diamond operator. In general, $\mathbf{a}_n$ (where $n \geq 1$) refers back to the world bound by the nth modal operator that takes scope over it, counting from (structural) top to bottom (if there is such an operator), while $\mathbf{a}_0$ always refers back to the outermost world. As indicated above (with (11) under a definition like (9)), context-dependent **A**ctually, i.e. that operator which always picks the actual world, has to be kept distinct from $\mathbf{a}_0$ and will be ignored for the time being.

It is known that such $\mathbf{a}_n$-operators do not increase the expressive power of M_0 (cf. Yanovich 2015, in a system utilizing (10) instead of (9)). In particular, (17) can be equivalently expressed in M_0:

(18) $\quad \Diamond (\psi \wedge \Diamond \varphi)$

In (18), the formula ψ of (17) has been 'raised' to a position where it achieves the intended interpretive effect, i.e. below the outermost, but above the second $\Diamond$ in (17), while φ remains below the second one to be evaluated at w_k. Such semantically pertinent transformations that re-position the scopes can be found to ultimately eliminate any backwards-looking operators within (extended) M_0-formulae.[14] However, the redundancy of $\mathbf{a}_n$-operators is lost by passing from propositional M_0 to *Quantified Modal Logic* M_1 with the usual interpretation of predicates, individual variables and a universal quantifier over individuals. (19) is a case in point:

(19) $\quad \Diamond (\forall x)[\mathbf{a}_0(Q(x)) \rightarrow Q(x)]$

For illustrative purposes, assume that $\Diamond$ again shifts the evaluation of its scope from w_i to w_j. The leftmost predication $Q(x)$ still is interpreted at w_i, due to $\mathbf{a}_0$, whereas the second $Q(x)$ is evaluated at w_j. But this time, contrary to the examples above, there is no equivalent M_1-formula. In particular, the individual quantifier $\forall x$ right above the $\mathbf{a}_0$-modified predicate blocks transformations that raise the first $Q(x)$ to the position where the $\mathbf{a}_0$-operator becomes redundant. This is simply due to the fact that $\mathbf{a}_0(Q(x))$ has to stay in the scope of its quantifier.[15]

---

[14] See Yanovich (2015) for a recent formulation of such a procedure.
[15] Of course, the fact that the usual M_0-transformations fail does not mean that there is no other way of expressing (19) in M_1. But see Hodes (1984), Wehmeier (2003), and Yanovich (2015) for proofs.

Backwards-looking operators have been argued to be needed in order to account for the truth conditions of natural language sentences like the following from Saarinen (1979, 343):

(20) Every man who ever supported the Vietnam War will have to admit that now he believes that he was an idiot then.

The point of (20) is that it has readings according to which every man who at any time in the past (contributed by *ever*) supported the Vietnam War has to admit at some time in the future that he believes at the time of utterance (contributed by *now*) that he was an idiot at that time in the past days when he used to support the Vietnam War. That being an idiot and supporting the Vietnam War are evaluated with respect to the same temporal index is the contribution of *then*, which thus seems to spell out an instance of ϑ_n overtly. The following formulae bring out the readings in questions:[16]

(21) a. $(\forall x)[\mathbf{M}_{w,t}(x) \to (\forall t' < t)[\mathbf{S}_{w,t'}(x) \to$
$(\exists t'' > t)[(\forall w')[\mathbf{A}_{w,t''}(x)(w') \to (\forall w'')[\mathbf{B}_{w',t_c}(x)(w'') \to \mathbf{I}_{w'',t'}(x)]]]]]$
b. $(\forall x)[\mathbf{M}_{w,t}(x) \to (\forall t' < t)[\mathbf{S}_{w,t'}(x) \to$
$(\exists t'' > t')[(\forall w')[\mathbf{A}_{w,t''}(x)(w') \to (\forall w'')[\mathbf{B}_{w',t_c}(x)(w'') \to \mathbf{I}_{w'',t'}(x)]]]]]$
c. $(\forall x)[\mathbf{M}_{w,t}(x) \to (\forall t' < t)[\mathbf{S}_{w,t'}(x) \to$
$(\exists t'' > t_c)[(\forall w')[\mathbf{A}_{w,t''}(x)(w') \to (\forall w'')[\mathbf{B}_{w',t_c}(x)(w'') \to \mathbf{I}_{w'',t'}(x)]]]]]$

These formulae differ only in the time with respect to which the future tense of *have to admit* is evaluated. *t* represents some index-time not necessarily identical to the time of utterance, which is denoted by t_c. Note that *t* makes its reappearance as the time from which the time contributed by *ever* is calculated, i.e. in the formalization of the relative clause. This is in line with Keshet's (2008) Intersective Predicate Generalization.[17]

2.2. Syntax and semantics of (tensed) intensional type logic

In this section we define a language *IL* that slightly extends Montague's (1973) tensed variant of his (1970) system by allowing for context

---

[16] Following Hintikka (1969), the function $\mathbf{B}_{w,t}(x)$ characterizes the doxastic perspective of an individual *x* in the world *w* at the time *t* by assigning truth to all worlds *w'* that are compatible with what *x* believes in *w* at *t*. Similarly for $\mathbf{A}_{w,t}(x)$. Below, we substitute **B** for **A** just for convenience.

[17] A reviewer pointed out to us that (21a) and (21b) are unlikely readings of (20). We still bring them up for later reference: in Section 2.4 we will use (21a–c) to elucidate a general strategy of dealing with such formulae in terms of higher-order intensional logic.

dependence in terms of double indexing. *IL* is based on the following set of *intensional types*:

(22) *Intensional Types*
Let e, t, and s be some fixed distinct objects. Then T_{IL} is the smallest set satisfying the following conditions:
a. $t \in T_{IL}$; $e \in T_{IL}$;
b. if $\sigma, \tau \in T_{IL}$, then $\langle \sigma, \tau \rangle \in T_{IL}$;
c. if $\tau \in T_{IL}$, then $\langle s, \tau \rangle \in T_{IL}$.

T_{IL} is a subset of the set T_2 of *two-sorted types*, which comprise e, t, and s and are closed under pairing. The difference is that the label s, which stands for the sort of indices, is not a full type in *IL* but is only used to mark intensions. Hence there are no *IL*-terms of types s, $\langle s, s \rangle$, or of the form $\langle \langle \tau, s \rangle, \tau \rangle$ while expressions of the types $\langle s, \langle s, t \rangle \rangle$, $\langle s, \langle s, \tau \rangle, \langle \langle s, \tau \rangle, t \rangle \rangle$, etc. do exist (provided that $\tau \in T_{IL}$, of course). This restriction on the realm of types might be taken to induce a restriction of the expressive power of *IL* in comparison to full *Ty2*. However, it is actually less severe than it seems (cf. Zimmermann 1989). Thus, as will be demonstrated below, even though *IL* is a sublanguage of *Ty2*, the relevant part that is exclusive to the latter can be coded in the former.

Given the two sets of basic terms in (23), the terms of *IL* are defined as in (24):

(23) Basic Terms of *IL*
a. Con_τ is the (denumerably infinite) set of all non-logical constants of type $\tau \in T_{IL}$;
b. Var_τ is the (denumerably infinite) set of all variables of type $\tau \in T_{IL}$.

(24) The family $(IL_\tau)_{\tau \in T_{IL}}$ of *IL*-terms
a. If $c \in Con_\tau$, then $c \in IL_\tau$;
b. If $x \in Var_\tau$, then $x \in IL_\tau$;
c. If $\alpha, \beta \in IL_\tau$, then $[\alpha = \beta] \in IL_t$;
d. If $x \in Var_\sigma$ and $\alpha \in IL_\tau$, then $[\lambda x.\alpha] \in IL_{\langle \sigma, \tau \rangle}$;
e. If $\alpha \in IL_{\langle \sigma, \tau \rangle}$ and $\beta \in IL_\sigma$, then $\alpha(\beta) \in IL_\tau$;
f. If $\alpha \in IL_\tau$, then $[^\wedge \alpha] \in IL_{\langle s, \tau \rangle}$;
g. If $\alpha \in IL_{\langle s, \tau \rangle}$ then $[^\vee \alpha] \in IL_\tau$.

(24a)–(24e) are familiar from (extensional) type logic; cf. Church (1940). The operators $^\wedge$ and $^\vee$ indicate abstraction from and application to the implicit index parameter, respectively. In this respect, *IL* differs from *Ty2*, where *s* is a full type (along *e* and *t*) and thus, informally, what is denoted by $^\wedge\alpha$ in *IL* can be expressed by $\lambda i_0.\alpha$ (see Gallin's 1975 translation-procedure in the next section). In order to interpret the typed *IL*-terms, the domains of denotation have to be fixed:

(25) a. $D_e = D$, a (fixed and non-empty) set of (possible) individuals
 b. $D_t = \{0,1\}$
 c. $D_{\langle \sigma, \tau \rangle} = D_\tau^{D_\sigma}$
 d. $D_{\langle s, \tau \rangle} = D_\tau^{(W \times T)}$, where *W* and *T* are (fixed and non-empty) sets of worlds and times.

IL-terms receive their denotations relative to *IL*-models $M = (D,T,W,F_M)$, variable assignments *g*, and points of reference $((w_c,t_c),(w,t))$, where:

- $F_M(\mathbf{c}): (W \times T) \to D_\tau$, for any $\tau \in T_{IL}$ and $\mathbf{c} \in \text{Con}_\tau$;
- $g(x) \in D_\tau$, for any $\tau \in T_{IL}$ and $x \in \text{Var}_\tau$.

(26) Interpretation of *IL*-terms
 Let *g* be a variable assignment, $M = (D,T,W,F_M)$ an *IL*-model, $w_c,w \in W$, and $t_c,t \in T$. Then, for any type τ and any $\alpha \in IL_\tau$, the denotation $[\![\alpha]\!]^{M,(w_c,t_c),(w,t),g} \in D_\tau$ of α is defined by the following recursion:
 a. $[\![\mathbf{c}]\!]^{M,(w_c,t_c),(w,t),g} = F(\mathbf{c})(w,t)$, if $\mathbf{c} \in \text{Con}_\tau$;
 b. $[\![x]\!]^{M,(w_c,t_c),(w,t),g} = g(x)$ if $x \in \text{Var}_\tau$;
 c. $[\![[\alpha = \beta]]\!]^{M,(w_c,t_c),(w,t),g} = 1$ iff
 $[\![\alpha]\!]^{M,(w_c,t_c),(w,t),g} = [\![\beta]\!]^{M,(w_c,t_c),(w,t),g}$;
 d. $[\![[\lambda x.\alpha]]\!]^{M,(w_c,t_c),(w,t),g}(u) = [\![\alpha]\!]^{M,(w_c,t_c),(w,t),g[x/u]}$, for any $u \in D_\sigma$ (where $x \in \text{Var}_\sigma$);[18]
 e. $[\![\alpha(\beta)]\!]^{M,(w_c,t_c),(w,t),g} = [\![\alpha]\!]^{M,(w_c,t_c),(w,t),g}([\![\beta]\!]^{M,(w_c,t_c),(w,t),g})$;
 f. $[\![[^\wedge(\alpha)]]\!]^{M,(w_c,t_c),(w,t),g}(w',t') = [\![\alpha]\!]^{M,(w_c,t_c),(w',t'),g}$, for any $w' \in W$ and $t' \in T$;
 g. $[\![[^\vee(\alpha)]]\!]^{M,(w_c,t_c),(w,t),g} = [\![\alpha]\!]^{M,(w_c,t_c),(w,t),g}(w,t)$.

---

[18] As usual, $g[x/u] = [g \setminus \{(x,g(x))\}] \cup \{(x,u)\}$. – It should be noted that (26d) is an implicit (or *pointwise*) definition of the denotation of the λ-term, which is that function with domain D_σ that satisfies the equations. Analogous remarks apply to (26f) below

A few details of the interpretation are noteworthy. To begin with, the double indexing plays no rôle so far but has only been made to prepare the ground for context-sensitive operators like **A**ctually and **N**ow to be defined later. Secondly, there is an interpretational asymmetry in the basic terms: the extensions of constants depend on the index parameter, those of variables do not; hence only constants may be construed as carrying about an implicit index variable and will be treated as such in the standard translation to be provided. Finally, the expressive means of *IL* are more powerful than it may appear since the usual logical operators (sentential connectives, the universal quantifier $\forall x$ as well as the necessity operator $\Box$) can be defined on the basis of those provided by (26):[19]

(27) For any type $\tau \in T_{IL}$:
   a. $(\forall x)\varphi$ is short for $[(\lambda x.\varphi) = (\lambda x.[x = x])]$
   for any $\varphi \in IL_t$ and $x \in Var_\tau$;
   b. $\neg\varphi$ is short for $[\varphi = (\forall x)x]$, where $x \in Var_t$;
   c. $[\varphi \wedge \psi]$ is short for $(\forall f)[\varphi = [f(\varphi) = f(\psi)]]$,
   where $\varphi, \psi \in IL_t$, and $f \in Var_{\langle t,t \rangle}$, but $f \notin Fr(\varphi) \cup Fr(\psi)$;
   d. $\Box\varphi$ is short for $([^\wedge\varphi] = [^\wedge(\forall x)[x = x]])$, where $\varphi \in IL_t$.

It should be noted that the (common) notation in (27d) obscures the fact that $\Box$ operates on a *proposition* obtained by applying $^\wedge$ to its 'prejacent' φ. These hidden occurrences of $^\wedge$-operators need to be kept track of when it comes to looking backwards in Section 2.5. As in Montague (1973), the box operator unrestrictedly quantifies over indices. In the current version of *IL*, operators quantifying separately over times and worlds are defined in terms of four particular constants $\sim_w$, $\sim_t$, $\prec$, and $\succ$ of type $\langle s,t \rangle$ that denote certain accessibility relations between indices in any *intended IL-model* $M = (D,T,W,F_M)$:

(28) a. $F_M(\prec)(w,t)(w',t') = 1$ iff $w = w'$ and $t' < t$;
   b. $F_M(\succ)(w,t)(w',t') = 1$ iff $w = w'$ and $t < t'$;
   c. $F_M(\sim_t)(w,t)(w',t') = 1$ iff $w = w'$;
   d. $F_M(\sim_w)(w,t)(w',t') = 1$ iff $t = t'$.

---

[19] The definitions in (27) follow Montague (1970, 387), who credits Tarski (1923) for (27b) and (27c). The restriction that *f* is not allowed to have free occurrences in φ and ψ is missing from Montague's formulation. Thanks to one of the reviewers for pointing out its need; see also Link and Varga von Kibéd (1975, 274).

Here '<' is taken to be a linear order modeling temporal precedence. The definitions in (28) take advantage of the fact that propositional constants are assigned intensions and denote propositions, thus relating the index of evaluation to the indices in their extension, which is the proposition denoted. In particular, < denotes the set of indices in the past of the current one, and similarly for >. $\sim_w$ generalizes a notation familiar from the interpretation of variable binding in that it denotes a relation between two indices that coincide except possibly in their second component. The constants in (28) may be used to define restricted quantification over indices in the style of (13b). As is usual in modal and temporal logic (but not necessary in *IL*), they are given a syncategorematic treatment as preceding arbitrary terms $\varphi \in IL_t$:

(29) Modal and temporal operators in *IL*
    a.   **L**φ is short for $[\lambda p.\Box[[^\vee p] \to \varphi]](\sim_w)$; $[p \in Var_{\langle s,t\rangle}]$
    b.   **P**φ is short for $[\lambda p.\Diamond [[^\vee p] \wedge \varphi]](\prec)$;
    c.   **F**φ is short for $[\lambda p.\Diamond [[^\vee p] \wedge \varphi]](\succ)$.

On the basis of these definitions, it is readily seen that **L** is an unrestricted universal quantifier over the world component of the index and **P**[ast] and **F**[uture] receive their standard tense-logical interpretations.[20]

Apart from the accessibilities defined in (28), we will avail ourselves to further *IL*-constants of type $\langle\langle s,t\rangle,\langle e,t\rangle\rangle$ to express propositional attitudes in the spirit of Hintikka (1969). We illustrate the strategy by one standard example, where **b** is a constant of type $\langle e,\langle s,t\rangle\rangle$ that represents the doxastic perspective of its outermost argument:

(30)   **B**$_\alpha\varphi$ is short for $[\lambda p.\Box[[^\vee p] \to \varphi]](\mathbf{b}(\alpha))$, where $\alpha \in IL_e$.

Under the assumption that doxastic alternatives are tenseless points in logical space, the interpretation function F_M of any intended *IL*-model would have to satisfy the following 'meaning postulate' (where *u* is an individual, *w* and *w'* are arbitrary worlds, and *t*, *t'* and *t''* are times in *M*):

---

[20] $\succ$ has only been introduced to emphasize the parallelism between the three operators in (29); **F** could also have been defined in terms of $\prec$, e.g. by the cumbersome term:
    (i) $(\lambda p.\Diamond [[^\vee p] \wedge \varphi])((\lambda O.[^\wedge O(\prec)])(\lambda p.[^\vee p]))$, where $O \in Var_{\langle\langle s,t\rangle,t\rangle}$.
Readers unfamiliar with *IL* should be warned that β-conversion is heavily restricted in that language: see Gallin's (1975) (correct) axiom schema AS4. In particular, it does not apply to the constellations in (29).

(31) $F_M(\mathbf{b})(u)(w,t)(w',t') = F_M(\mathbf{b})(u)(w,t)(w',t'')$

Since the operators in (29) and (30) are defined in terms of $\square$, they too introduce one implicit occurrence of $\wedge$ each; again, this will have to be kept in mind for the treatment of backwards-looking operators in Section 2.5.

Let us finally indicate how the two-dimensionality of our *IL*-version can be put to use to model context-dependence:

(32) Indexical Operators
   a. $[\![\mathbf{A}\varphi]\!]^{M,(w_c,t_c),(w,t),g} = 1$ iff $[\![\varphi]\!]^{M,(w_c,t_c),(w_c,t),g} = 1$;
   b. $[\![\mathbf{N}\varphi]\!]^{M,(w_c,t_c),(w,t),g} = 1$ iff $[\![\varphi]\!]^{M,(w_c,t_c),(w,t_c),g} = 1$.

2.3. From IL to Ty2

We will now assign to each *IL*-term α a term $\overline{\alpha}$ of the same type in which context and index are, respectively, represented by separate variables of type s, c and i_0. The terms $\overline{\alpha}$ derive from the language *Ty2* of *two-sorted type theory*, which is just like *IL*, except that it treats s as a second sort alongside e, which freely combines with other types, all of which (including s itself) host countably many constants and variables. However, unlike *IL*, the interpretation functions F_M of *Ty2*-models $M = (D,T,W,F_M)$ assign appropriate extensions rather than intensions to constants $\mathbf{c} \in Con_\tau$: $F_M(\mathbf{c}) \in D_\tau$, whenever $\tau \in T_2$ (= the set of all types generated from e, s, and t by pairing). To make up for this difference between source and target language, we follow Gallin's (1975) (p. 61) standard translation (33) and take *IL*-constants of types $\tau \in T_{IL}$ to also feature as *Ty2*-constants of type $\langle s,\tau \rangle$ that drag the index along when they get translated.

(33) Standard translation of *IL* to *Ty2*
   a. $\overline{\mathbf{c}} = \mathbf{c}(i_0)$, for constants $\mathbf{c}$;
   b. $\overline{x} = x$, for variables x;
   c. $\overline{[\alpha = \beta]} = [\overline{\alpha} = \overline{\beta}]$;
   d. $\overline{\alpha(\beta)} = \overline{\alpha}(\overline{\beta})$;
   e. $\overline{[\lambda x.\alpha]} = [\lambda x.\overline{\alpha}]$;
   f. $\overline{[^\wedge \alpha]} = [\lambda i_0.\overline{\alpha}]$;
   g. $\overline{[^\vee \alpha]} = \overline{\alpha}(i_0)$.

Since the clause (33a) covers the constants $\sim_w$, $<$, $>$, and $\mathbf{b}$, there is no need to define the translations for the operators introduced in (29) and (30), whose denotations come out as expected:

(34) If $\varphi \in IL_t$, $M = (D,T,W,F_M)$ is an (intended) Ty2-model, g is a (suitable) Ty2-assignment and $g(x) = (w,t)$, then:
  a. $[\![\mathbf{L}\varphi]\!]^{M,g} = 1$
  iff $[\![(\forall i_1)[\sim_\mathbf{w}(i_0)(i_1) \to \overline{\varphi}[i_0/i_1]]]\!]^{M,g} = 1$
  iff $[\![\overline{\varphi}]\!]^{M,g[i_0/(w',t)]} = 1$, for any $w' \in W$
  b. $[\![\mathbf{P}\varphi]\!]^{M,g} = 1$
  iff $[\![(\exists i_1)[\prec(i_0)(i_1) \land \overline{\varphi}[i_0/i_1]]]\!]^{M,g} = 1$
  iff $[\![\overline{\varphi}]\!]^{M,g[i_0/(w,t')]} = 1$, for some $t' < t$
  c. $[\![\mathbf{F}\varphi]\!]^{M,g} = 1$
  iff $[\![(\exists i_1)[\succ(i_0)(i_1) \land \overline{\varphi}[i_0/i_1]]]\!]^{M,g} = 1$
  iff $[\![\overline{\varphi}]\!]^{M,g[i_0/(w,t')]} = 1$, for some $t' > t$
  d. $[\![\mathbf{B}_x\varphi]\!]^{M,g} = 1$
  iff $[\![(\forall i_1)[\mathbf{b}(x)(i_0)(i_1) \to \overline{\varphi}[i_0/i_1]]]\!]^{M,g} = 1$
  iff $[\![\overline{\varphi}]\!]^{M,g[i_0/(w',t)]} = 1$, for any w' in x's doxastic alternatives in w

Finally, to treat two-dimensionality at all points of reference, the context-dependent operators **A** and **N** are translated by invoking a fixed variable c of type s, which is distinct from all variables i_n:[21]

(35) a. $\overline{\mathbf{A}\varphi} = (\exists i_1)[\sim_\mathbf{w}(i_1)(i_0) \land \sim_\mathbf{t}(i_1)(c) \land [\lambda i_0.\overline{\varphi}](i_1)]$;
  b. $\overline{\mathbf{N}\varphi} = (\exists i_1)[\sim_\mathbf{w}(i_1)(c) \land \sim_\mathbf{t}(i_1)(i_0) \land [\lambda i_0.\overline{\varphi}](i_1)]$.

Gallin's standard translation (33) preserves the local denotations of IL-terms in the following sense:

(36) *Lemma* (Gallin 1975, 62)
  Let $M = (D,T,W,F_M)$ be an IL-model, g an appropriate assignment, $((w_c,t_c),(w,t))$ a point of reference (in M), and $M^* = (D,T,W,F_{M^*})$ and g^* a Ty2-model and assignment such that:
  $F_M \subseteq F_{M^*}$, $g \subseteq g^*$, $g^*(c) = (w_c,t_c)$, and $g^*(i_0) = (w,t)$.
  Then for any $\tau \in TIL$ and $\alpha \in IL_\tau$:
  $[\![\alpha]\!]^{M,(w_c,t_c),(w,t),g} = [\![\overline{\alpha}]\!]^{M^*,g^*}$.

---

[21] Since in any intended model, any index satisfying the first two conjuncts in the scope of '$\exists i_1$' is uniquely determined, the translations given in (35) could also have been given in terms of universal quantification. – A reviewer rightly pointed out that – contrary to our above announcement – the definitions in (35) make use of more than two variables of type s. However, given that i_1 does not occur freely in $[\lambda i_0.\overline{\varphi}]$, the *definiens* in (35a) is equivalent to:
  (i). $[\lambda P.(\exists i_0)[P(\sim_\mathbf{w}(i_0)) \land \sim_\mathbf{t}(i_0)(c) \land [\lambda i_0.\overline{\varphi}](i_0)]](\lambda p.p(i_0))$
where P and p are variables of type $\langle\langle s,t\rangle,t\rangle$ and $\langle s,t\rangle$, respectively. Similarly for (35b).

2.4. Examples revisited

Returning to the crucial examples, repeated from above, it is easy to show that *IL* is able to circumvent the problems of M_1. This is due to its higher order, which allows abstraction from expressions of arbitrary complexity and of arbitrary type (in T_{IL}). It is thus possible to 'strand' bound (individual) variables in the position they must stay in and abstract from the whole predicate below the outermost ◇-operator. But since it would then be outside the scope of the lower ◇-operator, there is no need for $∃_0$ anymore. It becomes as redundant as in the initial *ML* examples. *IL*'s (38) expresses the same truth conditions as *Ty2*'s counterpart of (19), (37), which can be checked by translating (38) into *Ty2* with the help of (33)–(35), where it then can be reduced to (37) (modulo the names of bound variables):[22]

(19)   ◇ $(\forall x)[∃_0(Q(x)) \to Q(x)]$
(37) $(\exists i_1)[\sim_w (i_1)(i_0) \land (\forall x)[\mathbf{Q}_{i_0}(x) \to \mathbf{Q}_{i_1}(x)]]$
(38) $[\lambda \mathcal{R}. \diamond (\forall x)[\mathcal{R}(x) \to \mathbf{Q}(x)]](\mathbf{Q})$ $\mathcal{R} \in Var_{(e,t)}$

The rationale behind the construction of this formula is the same as in (18). Every predicate that is not evaluated against the temporal index introduced by the closest temporal operator is 'raised' in such a way that it ends up in the scope of the right operator. This time this is made possible by higher order abstraction. That this is no accident but part of a general strategy can be seen by considering *IL*'s account of the more involved natural language example, repeated from above:

(20)   Every man who ever supported the Vietnam War will have to admit that now he believes that he was an idiot then.

The most interesting reading is expressed by the following formula:

(21a) $(\forall x)[\mathbf{M}_{w,t}(x) \to (\forall t' < t)[\mathbf{S}_{w,t'}(x) \to (\exists t'' > t)[(\forall w')[\mathbf{A}_{w,t''}(x)(w') \to$
         $(\forall w'')[\mathbf{B}_{w',t_c}(x)(w'') \to \mathbf{I}_{w'',t'}(x)]]]]]$

The following *IL* formula indeed expresses exactly the truth conditions of (21a) without any help from backwards-looking operators. This can be

---

[22](38) is one of the counter-examples to β-conversion alluded to in fn. 20 above: substituting $\mathcal{R}$ by the 'modally open' argument would lead to an illicit binding of the implicit index-variable in the predicate constant **Q** by the ◇-operator.

seen by applying the translation function $\overline{\bullet}$ to the whole formula and reducing it further ($\mathbf{H}\varphi$ is short for $\neg\mathbf{P}\neg\varphi$, and we substitute *believe* for *admit* for convenience).

(39) $[\lambda\mathcal{O}_2.[\lambda\mathcal{O}_1.(\forall x)[\mathbf{M}(x) \to \mathcal{O}_1(^\wedge\mathbf{S}(x) \to \mathcal{O}_1,\mathcal{O}_2 \in Var_{\langle\langle s,t\rangle,t\rangle}$-
$[\lambda q.\mathcal{O}_2(^\wedge\mathbf{B}_x(\mathbf{N}(\mathbf{B}_x(^\vee q \wedge \mathbf{I}(x)))))](\sim_\mathbf{t}))]](\lambda p.\mathbf{H}^\vee p)](\lambda q.\mathbf{F}^\vee q)$

Thus, *IL* seems to be able simulate to backwards-looking operators by making use of higher types. This also holds for the other readings of (20) listed above. The following *IL* formulae do the trick as the inclined reader is invited to verify:

(21b) $(\forall x)[\mathbf{M}_{w,t}(x) \to (\forall t' < t)[\mathbf{S}_{w,t'}(x) \to (\exists t'' > t')[(\forall w')[\mathbf{A}_{w,t''}(x)(w') \to$
$(\forall w'')[\mathbf{B}_{w',t_c}(x)(w'') \to \mathbf{I}_{w'',t'}(x)]]]]]$

(40) $[\lambda\mathcal{O}.(\forall x)[\mathbf{M}(x) \to \mathcal{O}(^\wedge\mathbf{S}(x) \to$
$[\lambda q.\mathbf{FB}_x(\mathbf{N}(\mathbf{B}_x(^\vee q \wedge \mathbf{I}(x)))))](\sim_\mathbf{t}))]](\lambda p.\mathbf{H}^\vee p)$ $\mathcal{O} \in Var_{\langle\langle s,t\rangle,t\rangle}$

(21c) $(\forall x)[\mathbf{M}_{w,t}(x) \to (\forall t' < t)[\mathbf{S}_{w,t'}(x) \to (\exists t'' > t_c)[(\forall w')[\mathbf{A}_{w,t''}(x)(w') \to$
$(\forall w'')[\mathbf{B}_{w',t_c}(x)(w'') \to \mathbf{I}_{w'',t'}(x)]]]]]$

(41) $[\lambda\mathcal{O}.(\forall x)[\mathbf{M}(x) \to \mathcal{O}(^\wedge\mathbf{S}(x) \to$
$[\lambda q.\mathbf{NFB}_x(\mathbf{N}(\mathbf{B}_x(^\vee q \wedge \mathbf{I}(x)))))](\sim_\mathbf{t}))]](\lambda p.\mathbf{H}^\vee p)$

2.5. Implementing backwards-looking operators in IL

This section identifies the general mechanism behind the observations of the previous section. To this end, *IL* is endowed with a backwards-looking operator. Then an argument is given why this addition doesn't increase the expressive power of the original language.

The following type-theoretic implementation is in the spirit of Yanovich's (2015) strategy, which is why the resulting language is called '*YIL*'. The syntax is straightforward and thus omitted. The only innovation with respect to (24) above are the *backwards-looking* operators $\mathbf{\partial}_r^l$ (where l and r are natural numbers), which combine with terms of type $\langle s,t\rangle$ to produce terms of type t.

The semantics replaces the index in (26) by two parameters: a denumerable sequence ρ of indices and a natural number n. In view of ρ, *YIL* may be regarded as a multi-dimensional version of *IL*, although the rôle of the additional dimensions should not be confused with that of the context, which continues to be the relevant parameter for indexical operators. Rather, the function of ρ is to keep apart various occurrences of the ^-operator (mostly hidden in boldface constants). The other new parameter is a running index that is increased by any occurrence of $^\wedge$; hence n counts the number of intensional operators under which a term is embedded. At the same time, it indicates the local point of evaluation, as can be gleaned from the clause (42b) dealing with constants. Finally, the backwards-looking operators $Ⅎ_r^l$ tamper with the second rôle of n in that they shift the local point of evaluation from $\rho(n)$ to $(\rho(l)_1, \rho(r)_2)$ without affecting the depth of embedding. This divergence from Yanovich's (2015) operators (which shift the whole index) reflects the two-component indices of our version of IL (as opposed to its two-dimensionality); in particular, it facilitates dealing with quantification over 'mixed' indices whose components have been introduced by separate operators, like the ones at which the predicate I[diot] is evaluated in (21).

(42) Interpretation of YIL-terms
For any type $\tau \in T_{IL}$
a. $[\![x]\!]^{M,(w_c,t_c),\rho,n,g} = g(x)$
b. $[\![\mathbf{c}]\!]^{M,(w_c,t_c),\rho,n,g} = F_M(\mathbf{c})(\rho(n))$
c. $[\![[\alpha = \beta]]\!]^{M,(w_c,t_c),\rho,n,g} = 1$ iff $[\![\alpha]\!]^{M,(w_c,t_c),\rho,n,g} = [\![\beta]\!]^{M,(w_c,t_c),\rho,n,g}$
d. $[\![\alpha(\beta)]\!]^{M,(w_c,t_c),\rho,n,g} = [\![\alpha]\!]^{M,(w_c,t_c),\rho,n,g}([\![\beta]\!]^{M,(w_c,t_c),\rho,n,g})$
e. $[\![[\lambda x.\alpha]]\!]^{M,(w_c,t_c),\rho,n,g}(u) = [\![\alpha]\!]^{M,(w_c,t_c),\rho,n,g[x/u]}$
f. $[\![[^\wedge(\alpha)]]\!]^{M,(w_c,t_c),\rho,n,g}(i) = [\![\alpha]\!]^{M,(w_c,t_c),\rho[n+1/i],n+1,g}$
g. $[\![[^\vee(\alpha)]]\!]^{M,(w_c,t_c),\rho,n,g} = [\![\alpha]\!]^{M,(w_c,t_c),\rho,n,g}(\rho(n))$
h. $[\![Ɐ_r^l\varphi]\!]^{M,(w_c,t_c),\rho,n,g} = [\![\varphi]\!]^{M,(w_c,t_c),\rho[n+1/(\rho(l)_1,\rho(r)_2)],n+1,g}$
i. $[\![\mathbf{N}\varphi]\!]^{M,(w_c,t_c),\rho,n,g} = [\![\varphi]\!]^{M,(w_c,t_c),\rho[n+1/(\rho(n)_1,t_c)],n+1,g}$
j. $[\![\mathbf{A}\varphi]\!]^{M,(w_c,t_c),\rho,n,g} = [\![\varphi]\!]^{M,(w_c,t_c),\rho[n+1/(w_c,\rho(n)_2)],n+1,g}$

It is easy to see that, according to (42), the following equations hold:

(43) a. $[\![Ɐ_r^l\varphi]\!]^{M,(w_c,t_c),\rho,n,g} = [\![^\wedge\varphi]\!]^{M,(w_c,t_c),\rho,n,g}(\rho(l)_1, \rho(r)_2)$
b. $[\![\mathbf{N}\varphi]\!]^{M,(w_c,t_c),\rho,n,g} = [\![^\wedge\varphi]\!]^{M,(w_c,t_c),\rho,n,g}(\rho(n)_1, t_c)$
c. $[\![\mathbf{A}\varphi]\!]^{M,(w_c,t_c),\rho,n,g} = [\![^\wedge\varphi]\!]^{M,(w_c,t_c),\rho,n,g}(w_c, \rho(n)_2)$

The equations in (43) suggest an alternative treatment of $Ɐ_r^l$, **N**, and **A** as operating on propositions obtained by $^\wedge$-abstraction. We have chosen a more conservative formalization in terms of formula-operators instead. However, it will be useful in what follows to think of these operators as introducing hidden occurrences of $^\wedge$-just like the modal and tense operators.

As will become apparent in due course, *YIL* is a conservative extension of *IL*: *YIL*-terms that do not contain any backwards-looking operators $ℶ_r^l$ come out as equivalent to *IL*-terms, in a sense to be made precise below (cf. Section 2.6). The reader may wonder why we followed Yanovich (2015) in restricting the application of backwards-looking operators to propositions, rather than having them operate on arbitrary intensional types. Such a generalization would certainly be natural given the type-theoretic setting; and it would also be desirable in view of various 'generalized *de re*' phenomena observed in the literature from Bäuerle (1983) onward. However, as far as we can see, the inclusion of non-propositional backwards-looking operators would require a translation of *YIL* into a language of *three-sorted* type theory (mentioned in Section 1, fn. 11 above) in which the specification of 'mixed' indices in (42h) could be made in all types. Similar remarks apply to the indexical operators **A** and **N**. The reason why we content ourselves with *Ty2* will become apparent at the end of this section. In any case, the operators in (42) suffice to treat Saarinen's (1979) critical example (20):[23]

(44) $(\forall x)[\mathbf{M}(x) \to \mathbf{H}(\mathbf{S}(x) \to ℶ_0^0(\mathbf{FB}_x(\mathbf{N}(\mathbf{B}_x(ℶ_1^6(\mathbf{I}(x)))))))]$ (≡ (39))
(45) $(\forall x)[\mathbf{M}(x) \to \mathbf{H}(\mathbf{S}(x) \to \mathbf{FB}_x(\mathbf{N}(\mathbf{B}_x(ℶ_1^5(\mathbf{I}(x))))))]$ (≡ (40))
(46) $(\forall x)[\mathbf{M}(x) \to \mathbf{H}(\mathbf{S}(x) \to \mathbf{NFB}_x(\mathbf{N}(\mathbf{B}_x(ℶ_1^6(\mathbf{I}(x))))))]$ (≡ (41))

In order to compare its expressivity to that of *IL*, we need to cut down *YIL*'s multi-dimensionality, which it owes to the sequence-of-indices parameter ρ in (42). We proceed by giving a syntactic characterization of the degree $\sigma(\alpha)$ of an *YIL*-term α's 'modal openness':[24]

(47)

α	$\sigma(\alpha)$
x	0
c	0
$[\beta = \gamma]$	$\max(\sigma(\beta),\sigma(\gamma))$
$\beta(\gamma)$	$\max(\sigma(\beta),\sigma(\gamma))$
$[\lambda x.\beta]$	$\sigma(\beta)$
$[^\wedge\beta]$	$\sigma(\beta) \dotminus 1$
$[^\vee\beta]$	$\sigma(\beta)$
$ℶ_r^l\beta$	$\max(\sigma(\beta),l,r)$
$\mathbf{N}\varphi$	$\sigma(\varphi)$
$\mathbf{A}\varphi$	$\sigma(\varphi)$

---

[23] A three-sorted treatment of backwards-looking operators would also have to distinguish (at least) two distinct versions of the ^- and ˅-operators.
[24] $n \dotminus m = \max(0, n-m)$ for any n,m; negative numbers are thus avoided.

It is readily seen that $\sigma(\alpha)$ gives the highest position in the ρ-sequence that α's denotation may depend on:

(48) *Lemma.* If $\rho(i) = \rho'(i)$ for all $0 \leq i \leq \sigma(\alpha)$, then:
$[\![\alpha]\!]^{M,(w_c,t_c),\rho,\sigma(\alpha),g} = [\![\alpha]\!]^{M,(w_c,t_c),\rho',\sigma(\alpha),g}$.

We leave the verification of (48) to the reader. Since the denotations of *IL*-terms only depend on one index (and the context), they can only be compared to *YIL*-tems whose denotation depends on the position 0 of parameter ρ. In view of (48), this means that we may concentrate on *YIL*-terms whose σ-value is 0.[25]

2.6. From YIL to IL

In the course of this section, we will show how *YIL*-terms translate into *IL*, thus simulating the interpretive effect of any backwards-looking operators by the expressive means of ordinary intensional type logic. To this end, we will first show that the translation Z_0 of *YIL* into *Ty2* satisfies the following:

(49) *Theorem*
Let M, M^*, g, g^* and (w_c,t_c) be as in (36), and let ρ be a sequence of indices (in $W \times T$) such that $\rho(k) = g^*(i_k)$ for any k.
a. If $\tau \in T_{IL}$, $\alpha \in YIL_\tau$, and $g^*(i_0) = g^*(c)$, then:
$[\![Z_0(\alpha)]\!]^{M^*,g^*} = [\![\alpha]\!]^{M,(w_c,t_c),\rho,0,g}$.
b. If $\varphi \in YIL_t$, then:
$[\![Z_0(\varphi)]\!]^{M^*,g^*} = [\![\varphi]\!]^{M,(w_c,t_c),\rho,0,g}$.

The first part of (49) says that the Z_0-translation of an unembedded *YIL*-term preserves its denotation *in any context*. Thus (49a) generalizes over terms *of all types*; but it is restricted in that denotational equivalence is confined to *diagonal points*. The second part of (49) says that the Z_0-translation of an unembedded *YIL*-term *of type t* preserves its truth value at any given point of reference. Thus (49b) generalizes over *all points of reference*; but it is restricted in that denotational equivalence is confined to *truth-valuable terms*. With these restrictions, (49) may be used to to show that the addition of ∂_r^i does not increase the expressive power of *IL*:

---

[25] The condition that $\sigma(\alpha) = 0$ thus corresponds to the semantic characterization of *sentences* given in Yanovich (2015, 72f., fn. 8).

(50) *Corollary*
   a. For any $\alpha \in \mathit{YIL}_\tau$ such that $\sigma(\alpha) = 0$, there is a $\gamma \in \mathit{IL}_\tau$ such that for any M, (w_c, t_c), g, and ρ:
   $[\![\alpha]\!]^{M,(w_c,t_c),\rho,0,g} = [\![\gamma]\!]^{M,(w_c,t_c),(w_c,t_c),g}$.
   b. For any $\varphi \in \mathit{YIL}_t$ there is a $\psi \in \mathit{IL}_t$ such that for any M, (w_c, t_c), (w,t), g, and ρ:
   $[\![\varphi]\!]^{M,(w_c,t_c),\rho,0,g} = [\![\psi]\!]^{M,(w_c,t_c),(w,t),g}$.

As already indicated, strengthening (49) and (50) so as to apply to arbitrary terms at arbitrary points of reference would require replacing *Ty2* with a three-sorted variant and *IL* with a more complex (and less standard) version of intensional type logic – which is why we leave it at that.

The subscript in the translation Z_0 in (49) refers to the variable marking the current index, which needs to be kept flexible when it comes to intensional embedding. We thus define a family $(Z_n)_{n\in\mathbb{N}}$:[26]

(51) *Translation*
   a. $Z_n(x) = x$
   b. $Z_n(\mathbf{c}) = \mathbf{c}(i_n)$
   c. $Z_n([\alpha = \beta]) = [Z_n(\alpha) = Z_n(\beta)]$
   d. $Z_n(\alpha(\beta)) = Z_n(\alpha)(Z_n(\beta))$
   e. $Z_n([\lambda x.\alpha]) = (\lambda x.Z_n(\alpha))$
   f. $Z_n([^\wedge \alpha]) = (\lambda i_{n+1}.Z_{n+1}(\alpha))$
   g. $Z_n([^\vee \alpha]) = Z_n(\alpha)(i_n)$
   h. $Z_n(\mathbf{B}_r^l \varphi) = [\lambda p.(\exists j)[\sim_\mathbf{w}(j)(i_l) \wedge \sim_\mathbf{t}(j)(i_r) \wedge p(j)]](Z_n(^\wedge \varphi))$
   $= [\lambda p.(\exists j)[\sim_\mathbf{w}(j)(i_l) \wedge \sim_\mathbf{t}(j)(i_r) \wedge p(j)]](\lambda i_{n+1}.Z_{n+1}(\varphi))$
   i. $Z_n(\mathbf{N}\varphi) = [\lambda p.(\exists j)[\sim_\mathbf{w}(j)(i_n) \wedge \sim_\mathbf{t}(j)(c) \wedge p(j)]](\lambda i_{n+1}.Z_{n+1}(\varphi))$
   j. $Z_n(\mathbf{A}\varphi) = [\lambda p.(\exists j)[\sim_\mathbf{w}(j)(c) \wedge \sim_\mathbf{t}(j)(i_n) \wedge p(j)]](\lambda i_{n+1}.Z_{n+1}(\varphi))$

(51) preserves the spirit of Gallin's (1975) standard translation (33) of *IL* into *Ty2* whilst taking our adaption (42) of Yanovich's (2015) treatment of looking backwards into account. In fact, as the reader may verify:[27]

---

[26] In (50h), $p \in \mathit{Var}_{(s,t)}$; the roundabout formulation merely avoids variable clashes in case (arbitrary but fixed) $i_k \in \mathit{Var}_s$ happen to be free in $Z_n(\varphi)$.

[27] (51) follows from the following two observations both of which are readily established by induction on α's structure:

   (i) If $i_m \in \mathit{Fr}(Z_n(\alpha))$, then $m = n$;
   (ii) $Z_n(\alpha) = Z_m(\alpha)[i_n/i_m]$.

   ...for any natural numbers n and m.

(52) $Z_0(\alpha) \equiv \overline{\alpha}$, for any *IL*-term a, where '$\equiv$' indicates logical equivalence.

The following correlation links the σ-value of an *IL*-term to the index variables in its Z_n-translation:

(53) *Proposition*
Let n and k be natural numbers such that $k \geq 1$. Then for any term $\alpha \in YIL_\tau$ (where $\tau \in T_{IL}$) the following holds:
If $n \geq \sigma(\alpha)$, then $i_{n+k} \notin Fr(Z_n(\alpha))$.

Disregarding *YIL*-terms a where $\sigma(\alpha) > 0$ (in line with Lemma (48)), (53) thus says that we may confine ourselves to their Z_0-translations, which only contain at most c and i_0 as free variables of type s: any other terms would call for additional parameters and thus be outside the expressive scope of *IL* by definition. Now the following observation is crucial:

(54) *Lemma*
Let M, M^*, g, g^* and (w_c, t_c) be as in (36), and let ρ be a sequence of indices (in $W \times T$) such that $\rho(k) = g^*(i_k)$ for any k. Then for any $\tau \in TIL$, $\alpha \in YIL_\tau$ and natural number n:
$[\![Z_0(\alpha)]\!]^{M^*,g^*} = [\![\alpha]\!]^{M,(w_c,t_c),\rho,n,g}$.

Together with the observation (52) above, (54) implies that the *YIL*-interpretation (42) conservatively extends the semantics (26) of *IL*, in the following sense:

(55) $[\![\alpha]\!]^{M,(w_c,t_c),\rho,0,g} = [\![\alpha]\!]^{M,(w_c,t_c),\rho(0),g}$, for any *IL*-term a.

To see that (55) holds, one may apply (54) to find that $[\![\alpha]\!]^{M,(w_c,t_c),\rho,0,g} = [\![Z_0(\alpha)]\!]^{M^*,g^*}$, which (by (52)) is the same as $[\![\overline{\alpha}]\!]^{M^*,g^*}$, which in turn is $[\![\alpha]\!]^{M,(w_c,t_c),g^*(i_0),g}$, by Gallin's (36); but then $g^*(i_0) = \rho(0)$, and (55) follows.

(54) can be established by a straightforward induction on the complexity of *YIL*-terms a, which we skip. And the special case of $n=0$ of (54) immediately implies both parts of (49): in (49a), context and index coincide in that $g^*(i_0) = g^*(c)$; and in (49b), $\tau = t$.

To see the consequences of the lemma for *IL*-expressibility, we will resort to the following:

(56)   *Theorem* (Zimmermann1989, 75)
   If $\beta \in Ty2_\tau$ meets conditions a.–c., then there is a $\gamma \in IL_\tau$ such that $\overline{\gamma}$
   is logically equivalent to β.[28]
   a. $\tau \in TIL$;
   b.   if **c** is a constant occurring in β, then **c** is a constant of some type $\langle s, \sigma \rangle$, where $\sigma \in TIL$;
   c.   if x is a variable occurring freely in β, then $x \in Var_\sigma \cup \{i_0\}$ where $\sigma \in TIL$.

With the aid of (56), both parts of (50) can now be derived. Ad a.: If $\alpha \in YIL_\tau$ and $\sigma(\alpha) = 0$, then $Z_0(\alpha)$ meets conditions (56a) and (56b) – because all Z_n-translations do, as a brief inspection of (51) shows. Moreover by (53), c and s_0 are the only free variables of type s in $Z_0(\alpha)$, and all other free variables in $Z_0(\alpha)$ are IL-variables. Hence the term $[\lambda c.Z_0(\alpha)](i_0)$ meets condition (56c), on top of (56a) and (56b). So by Theorem (56), it is equivalent to the standard translation $\overline{\gamma}$ of some IL-term y. But then, by Lemma (36), $[\![\gamma]\!]^{M,(w_c,t_c),(w_c,t_c),g} = [\![\overline{\gamma}]\!]^{M^*,g^*} = [\]\!]^{M^*,g^*} = [\![Z_0(\alpha)]\!]^{M^*,g^*[c/g^*(i_0)]} = [\![Z_0(\alpha)]\!]^{M^*,g^*} g^*(i_0) = g^*(c) = (w_c, t_c)$. Hence $[\![\gamma]\!]^{M,(w_c,t_c),(w_c,t_c),g} = [\![\alpha]\!]^{M,(w_c,t_c),\rho,0,g}$, by (49a). – Ad b.: In a similar vein, given $\varphi \in YIL_t$ with $\sigma(\varphi) = 0$, $[\lambda i_0.\lambda c.Z_0(\varphi)]$ meets the conditions (56a)–(56b) and is thus equivalent to the translation $\overline{\psi}$ of some $\psi \in IL_{\langle s, \langle s, t \rangle \rangle}$. As it turns out (and the reader may verify as an exercise), the IL-term $[\lambda p.\mathbf{AN}[\vee p]](\vee \psi)$ of type t then does the job.

Summing up, then, we have shown that backwards-looking operators do not increase the expressivity of intensional type logic on the level of term *denotation in* context; nor do they add expressive power on the level of sentential *truth at* a point of reference. A stronger result, concerning expressivity on the level of term denotation at a point of reference would have been obtained had we decided against index-splitting – or, presumably, if split indices had been made explicit in terms of three-sorted type theory.

3. Conclusions

The results in Section 2 may be seen as proof that the addition of backwards-looking operators does not increase the *truth-conditional* expressivity of the implicit parameter language of Intensional (Type) Logic. Still, the

---

[28]Here 'logical equivalence' is meant in the strong form that the denotations coincide in all models at all assignments.

backwards-looking operators have not been shown to be equivalent to corresponding simpler terms; rather, their elimination depends on their occurrence. For we have not given canonical translations of backwards-looking operators in *IL* and *Ty2*: although all of their occurrences can be paraphrased by some *IL* term, different paraphrases may be called for once they get further embedded. As a consequence, the above results do not directly bear on their application to the compositional semantics of natural language. As various semanticists have argued, the treatment of multiple intensional embedding in a compositional way calls for extensions of implicit parameter interpretation (e.g. by backwards-looking operators) or even full explicit systems.[29] According to these authors, what is at stake here is the fundamental assumption that there is an upper limit to the number of parameters denotations may depend on – one in possible world semantics, two in two-dimensional semantics: the additional parameters are needed when it comes to interpreting the constituents of pertinent multiple embeddings. So whereas we have shown that such multiple embeddings can be expressed in *IL*, we have not shown that so can their constituents, which actually seem to be outside the reach of any intensional approach, given their (potential) dependence on arbitrarily many parameters. Still, there is a way out here. As argued in detail in Zimmermann (2018), Frege's (1892) strategy of raising the level of sense parallel to the level of intensional embedding can be adapted to cope with multiple index-dependence, replacing the free index variables of explicit systems by bound ones, which are known to be expressible in *IL*. As it turns out, the 'sense-twisting' operators needed to deal with the examples discussed in Sections 2.1 and 2.4 along these lines, may be seen as an alternative route to looking backwards in intensional type logic.

Acknowledgments

Predecessors of this paper were circulated among our colleagues and presented at a colloquium meeting in Frankfurt in spring 2018. We would like to thank the readers and participants for a number of useful comments that saved us from various errors and omissions. In particular, we are indebted to Fabio Lampert, Kristina Liefke, and especially Kai Wehmeier and two anonymous reviewers for their extensive comments.

Disclosure statement

No potential conflict of interest was reported by the authors.

---

[29]See, e.g. Bäuerle (1983), Cresswell (1990), or Schlenker (2006).

Author's note

Only after the online publication it has been brought to our attention that Yanovich's (2015) paper contains a number of serious flaws; we are grateful to Philippe Schlenker, Amir Anvari, and particularly Hans Kamp, who provided us with his extensive/Comments on I. Yanoch's 'Expressive Power of "Now" and "Then"'/ (ms., 2017). Yet, although our own results heavily rely on constructions adapted from Yanovich's account, the proofs are quite independent from his and, as far as we can tell, do not inherit their inadequacies.

References

Bäuerle, R. 1983. "Pragmatisch-semantische Aspekte der NP-Interpretation." In *Allgemeine Sprachwissenschaft, Sprachtypologie und Textlinguistik*, edited by M. Faust, R. Harweg, W. Lehfeldt, and G. Wienold, 121–131. Tübingen: Narr

Church, A. 1940. "A Formulation of the Simple Theory of Types." *The Journal of Symbolic Logic* 5 (2): 56–68.

Cresswell, M. J. 1990. *Entities and Indices*. Dordrecht, Boston, London: Kluwer.

Dummett, M. 1981. *The Interpretation of Frege's Philosophy*. London: Duckworth.

Fine, K. 1975. "Some Connections Between Elementary and Modal Logic." *Studies in Logic and the Foundations of Mathematics* 82: 15–31.

Frege, G. 1892. "Über Sinn und Bedeutung." *Zeitschrift für Philosophie und philosophische Kritik* NF 100: 25–50.

Gallin, D. 1975. *Intensional and Higher-Order Modal Logic*. Amsterdam: North-Holland Pub. Company.

Goranko, V., and M. Otto. 2007. "Model Theory of Modal Logic." In *Handbook of Modal Logic*, edited by P. Blackburn, J. van Benthem, and F. Wolter, 249–329. Chap. 5. Amsterdam: Elsevier

Hazen, A. P., B. G. Rin, and K. F. Wehmeier. 2013. "Actuality in Propositional Modal Logic." *Studia Logica* 101 (3): 487–503.

Hintikka, J. 1969. "Semantics for Propositional Attitudes." In *Philosophical Logic*, edited by J. W. Davis, D. J. Hockney, and W. K. Wilson, 21–45. Dordrecht: Reidel.

Hodes, H. 1984. "Some Theorems on the Expressive Limitations of Modal Languages." *Journal of Philosophical Logic* 13: 13–26.

Kaplan, D. 1979. "On the Logic of Demonstratives." *Journal of Philosophical Logic* 8: 81–98.

Keshet, E. 2008. "Good Intensions: Paving Two Roads to a Theory of the *De re/De dicto* Distinction." PhD thesis, MIT.

Kripke, S. A. 1972. "Naming and Necessity." In *Semantics of Natural Language*, edited by D. Davidson and G. Harman, 253–355. Dordrecht: Reidel.

Lewis, D. K. 1980. "Index, Context, and Content." In *Philosophy and Grammar*, edited by S. Kanger and S. Öhman, 79–100. Dordrecht: Reidel.

Link, G., and M. Varga von Kibéd. 1975. "Review of Formal Philosophy. Selected Papers of Richard Montague." *Erkenntnis* 9 (2): 252–286.

Montague, R. 1970. "Universal Grammar." *Theoria* 36: 373–398.

Montague, R. 1973. "The Proper Treatment of Quantification in Ordinary English." In *Approaches to Natural Language*, edited by P. Suppes, J. Moravcsik, and J. Hintikka, 221–242. Dordrecht: Reidel.

Rabern, B. 2012. "Against the Identification of Assertoric Content with Compositional Value." *Synthese* 189: 75–96.

Saarinen, E. 1979. "Backwards-Looking Operators in Tense Logic and in Natural Language." In *Essays on Mathematical and Philosophical Logic*, edited by J. Hintikka, I. Niiniluoto, and E. Saarinen, 341–367. Dordrecht: Reidel.

Schlenker, P. 2006. "Ontological Symmetry in Language: A Brief Manifesto." *Mind & Language* 21 (4): 504–539.

Tarski, A. 1923. "O wyrazie pierwotnym logistyki." *Przegląd filozoficzny* 26: 66–89.

van Benthem, J. 2001. "Correspondence Theory." In *Handbook of Philosophical Logic* Vol. II, 2nd ed., edited by D. M. Gabbay and F. Guenthner, 321–408. Dordrecht: Kluwer.

Wehmeier, K. F. 2003. "World Travelling and Mood Swings." In *Foundations of the Formal Sciences II*, edited by B. Löwe, W. Malzkorn, and T. Räsch, 257–260. Dordrecht: Kluwer.

Yanovich, I. 2015. "Expressive Power of *Now* and *Then* Operators." *Journal of Logic, Language and Information* 24: 65–93.

Zimmermann, T. E. 1989. "Intensional Logic and Two-Sorted Type Theory." *The Journal of Symbolic Logic* 54 (1): 65–77.

Zimmermann, T. E. 2012. "Context Dependence." In *Handbook of Semantics* Vol. 3, edited by C. Maienborn, K. von Heusinger, and P. Portner, 2360–2407. Berlin and New York: de Gruyter.

Zimmermann, T. E. 2018. "Fregean Compositionality." In *The Science of Meaning*, edited by D. Ball and B. Rabern, 276–305. Oxford: Oxford University Press.

Index

Note: Page numbers followed by "n" denote endnotes.

accessibility relations 38, 39, 56n23
almost-explicit semantics 112
Aloni, M. 48n19
ambiguity 64–66
Anand, P. 77
anti-positing covert stuff arguments 129
arân (counterfactual) 67
Aristotle 1
assignatory type theory 13–16
assignment intension 9, 14
assignment-shifting treatments 3, 33, 44

backwards-looking operators 2, 3, 141, 147, 164; examples 147–149; *IL* to *Ty2* 154–155; implementation 157–160; in intensional type logic 147–163; revisiting examples 156–157; syntax and semantics of (tensed) 149–154; *YIL* to *IL* 160–163
Bäuerle, R. 159
van Benthem, J. 97, 143, 143n4
binding, variables 1, 27–38; counterpart semantics 39–44; epistemic contexts 44–54
Blackfoot 68, 75, 76–77, 82
Bliss, H. 68
Brogaard, B. 100n10

Cappelen, H. 111, 128
categorematic treatment 10, 12–13, 15
central coincidence 74, 75
Chalmers, D. J. 30, 57, 81
Chomsky, N. 61
Chung, S. 62
Church, A. 151
Cicero 53
Cohen, A. 3

coincidence 74
Collins, J. 122, 127–129, 131
compositional categorematic theory 130
compositional formalization 145
compositionality 9, 10, 17, 20
Context-Index Semantics 95, 95n4, 112
Context Semantics 105, 106, 112
Conventional Implicature (CI) 72–74
counterpart functions 47, 48
counterpart semantics 3, 33, 39–44, 50, 52; variables in 39–44
covert variables 3, 118–138; case against 127–129; case for 122–127
Cresswell, M. J. 50n20, 95, 100–106, 102n11, 103n11, 119, 124
Cumming, S. 45, 46, 48, 48n18

Davis, W. A. 81
Deal, A. R. 77–79
Déchaine, R. M. 68
Del Naja, Robert 31n3
denotation 18n22, 19, 130, 135, 141, 145–147, 151, 154, 155, 160, 163, 163n28, 164
de re epistemic modality 33n4, 41n10, 43, 46, 159
double indexing 144, 150, 152
Dowty, D. R. 124

Elbourne, P. D. 137
Enç, M. 96
English 18–20, 59, 64–68, 75, 77, 80–82, 90, 97, 122, 124, 134
epistemic contexts 32, 33, 39; variables in 44–54
epistemic modalities 30, 32, 43, 49, 56–57
eternalism 3, 95–101

INDEX

explicit semantics 3, 88, 91, 101, 111; *vs.* implicit semantics 112–114; *vs.* mixed semantics 106–112
explicit time variables 97
extensional functional application 17
extensional machinery 4, 87, 88, 95, 100, 106–110, 113
extensional operator 5, 18

von Fintel, K. 5n1, 17
formal semantics 3, 120, 121, 132, 136
Frantz, D. 76
Fregean extension 21, 24
Fregean type theory 2, 7, 20–24
Frege, G. 6, 9, 16, 17n19, 23n28, 111, 164
Frege's Conjecture 6, 10, 10n7, 12–13, 20, 21, 23
functional application 17, 18, 23
function composition 135
future particles 67

Gallin, D. 146, 146n12, 147, 155, 161, 162
Geach, P. T. 1
generalization 3, 76, 81, 82, 159
Gerbrandy, J. 48n19
Gerdts, D. 65, 66n9
Glanzberg, M. 104
Grimshaw, J. 60
Gunningham, Robin 31n3

Hale, K. 74, 75
Halkomelem 64–66, 65n7, 75, 80
Hawthorne, J. 111
Heim, I. 5n1, 10, 10n7, 17, 19n23, 72, 120
hēn (factual non-future) 67
Hintikka, J. 149n16, 153
Horn, L. R. 74
Hughes, G. E. 50n20

idiosyncratic properties 59
implicit semantics 107, 111; *vs.* explicit semantics 112–114
indexicality: quantificational approach and 74–82
index semantics 88–91, 89n1, 97, 105
INFL 3, 61, 61n4, 62, 64, 66, 67, 69, 75–76, 79–80, 82, 83; as indexical 74–75, 75n14
intensional logic 2, 3, 88, 146, 146n17, 147; backwards-looking operators in 147–163; examples 147–149; *IL* to *Ty2* 154–155; implementation 157–160; in intensional type logic 147–163; revisiting examples 156–157; syntax and semantics of (tensed) 149–154; *YIL* to *IL* 160–163
intensional operators 1, 24–25, 110, 158; assignatory type theory 13–16; Fregean type theory 2, 7; price categorematicity 17–20; quantifiers as 5–8; syncategorematic type theory 7–13
intensional system 89, 95, 97, 100; environments 140, 141; faithfulness 15; functional application 17; machinery 87, 88, 100, 107–110, 112, 113; types 150
interpretation 8, 12n10, 19, 24, 29, 31, 32, 35, 36, 40, 41, 56, 60, 64–66, 71, 73, 74, 77, 102, 133, 142, 144, 146, 148, 151–154, 158, 164
Intersective Predicate Generalization 149, 149n17

Jackendoff, R. 63
Jacobson, P. 1, 18n22, 131, 134–137

Kamp, H. 90, 95, 97, 124, 127
Kaplan, D. 46, 76, 81, 88, 90n2, 96, 100, 101, 104, 105, 110, 111, 143; on indexicals 91–95
Kaplan–Lewis operator argument 96, 98, 100, 103
Karttunen, L. 72, 73
kê (factual future) 67
Keshet, E. 149
King, J. 96, 96n7, 125
Kīsêdjê 66, 67, 71, 80
Köpping, J. 2
kôt (inferential future) 67
Kratzer, A. 10, 10n7, 19n23, 56n23, 102, 131, 134, 136, 137
Kripke, S. 27–29, 30n2, 31–33, 42–44, 44n13, 53, 56, 143
Kripke semantics 3, 41, 41n10, 50

Ladusaw, W. A. 62
lambda abstraction 6, 23n28
lambda abstractors 2, 16, 23, 24n30
Lapwai 78
Leibniz's law 42
Lepore, E. 128
Levin, H. 16n15
Lewis, D. 33, 39, 40, 40n9, 42–44, 81, 88, 89n1, 93–95, 95n4, 99, 105, 105n13, 110, 111, 113–114

linguistic evidence 59, 75, 102; location 63–66; person 67–69; possible world 66–67; time 60–63
linguistic semantics 25, 146
location: linguistic evidence 63–66; quantificational approach 71
logical space 142, 143

MacFarlane, J. 110
McKeever, M. 3, 110n17
McKinsey's axiom 143
man (witnessed) 67
metaphysical modality 3, 29, 30, 43
mixed semantics: *vs.* explicit semantics 106–112
modality 28
modal logic 144
modal propositional logic 141
model-theoretic semantics 40
Montague, R. 2, 145, 146n12, 147, 149, 152, 152n19; on pragmatic languages 88–91
Moss, S. 125

natural language 97, 102, 118, 145, 149; compositional semantics of 164; expressions of 119; semantics 3, 87–91, 106, 107
necessitarian 100–101, 103
necessitarianism 3, 101, 103, 106, 112
Nevins, A. 77
Nez Perce 77–80
Ninan, D. 48, 48n18, 49, 50, 50n20, 51n21, 52, 53n22, 96n7, 99n9
No INFL Without Indexicality principle (NIWI) 75–76, 80, 82
Nonato, R. 66, 67
non-central coincdicence 74, 75
non-future particles 67
null indexicals 81–82

objectual quantification 129
operator argument 93, 95–100

Partee, B. H. 96, 101, 119, 124
Percus, O. 101
person: linguistic evidence 67–69; quantificational approach 72–74
Peters, S. 72, 73
Phillips, E. 80
philosophical theory 59; location 63–66; person 67–69; possible world 66–67; time 60–63

Pickel, B. 46, 46n14
Pietroski, P. M. 137
positing covert variables 3, 118–138; case against 127–129; case for 122–127; misguide 129–132; objections 133–137
possible world: linguistic evidence 66–67; quantificational approach 71–72
Potts, C. 74
pragmatic languages: Montague on 88–91
Predicate Modification 10n7
presupposition 72, 73
price categorematicity 17–20
Principle of Vacuous Binding 37, 39
Prior, A. N. 1, 123
Priorean theory 124
proposition 3, 45, 72, 88, 90–92, 95, 96, 100, 101, 104, 105, 110–113, 141–145, 148, 152–153, 158, 159, 162
Pythagorus 41n10

quantificational approach 1, 4, 82–83; Blackfoot 76–77; English 81; Halkomelem 80; indexicality and 74–82; INFL as indexical 74–75; location 71; Nez Perce 77–80; no INFL without indexical 75–76; null indexicals 81–82; person 72–74; possible world 71–72; time 69–71
quantification theory 119–120
quantified epistemic modality 30, 32
quantified modal logic 27, 28, 30n1, 56–57; objectual interpretation of 29; standard semantics for 29–30
quantifier-like behaviour 118–120, 122–123
quantifiers 3, 5–7, 5n2, 9, 12, 25, 36, 37n6, 38, 39, 87, 98, 108; assignatory type theory 13–16; Fregean type theory 2, 7; as intensional operators 5–7; price categorematicity 17–20; symbol 34–35; syncategorematic type theory 7–13; tense theory 119–120, 122, 137
Quinean objections 27–28, 31
Quine, W. V. 1

Rabern, B. 3, 3n2, 14n13, 57, 104, 105
Recanati, F. 59, 59n1, 60–64, 70, 83, 90, 100n10
Richard, M. 95
Rini, A. A. 100, 102, 103n11
Ritter, E. 64, 68, 75, 76
Russell, N. 76
Russell, S. 80

Saarinen, E. 2n1, 149, 159
Santorio, P. 46, 46n15, 47, 48, 48n18
Schaffer, J. 3, 100, 102
Schlenker, P. 77, 91, 100, 102, 119
Schönfinkel, M. 1
Schroedinger equation 121, 133
Schwarz, W. 41n10, 50n20
Scott, D. 88, 89, 110
selectional restrictions 63
semantic argument 62–64, 66
semantic composition 9–10
semantic effects 126, 127, 136
semantic machinery 89, 92, 104, 105, 107
semantic parallelism 9, 102, 103, 114
semantic rules 17
semantics 136; behaviour 136–137; misguided 129–132; not misguided 132–133
Shift-Together principle 77, 78
Sommers, F. 1
de Sousa, M. S. C. 80
Speas, M. 101
Stalnaker, R. 95, 95n6, 101
Stanley, J. 110n18, 122, 126–128
Stone, M. 100–106
sui generis rule 21–23
symbolism 17n19
syncategorematic clauses 6
syncategorematic type theory 7–13
syntactic pollution 18, 19
syntactic-semantic architecture 137
Szabó, Z. G. 122, 126, 127

Tarski, A. 35, 35n5, 105n14, 152n19
temporal variables 3, 70, 120, 122, 125, 127–129, 133, 134
tense 60, 64, 66, 70, 75, 79, 118, 118n1; projection 102; quantifier theory of 119–120, 122, 137
three-sorted type theory 146n11, 159, 159n23
time: linguistic evidence 60–63; quantificational approach 69–71; variables 88

truth-conditional effects 72, 110n18, 123, 146, 147, 163
two-dimensional actuality operator 144
two-dimensional semantics 144, 164
two-sorted extensional type theory 146, 150, 154
Tyler, M. 82
type theory 2, 5n1, 6; assignatory type theory 13–16; extensional 6, 7; formulations of 6; syncategorematic type theory 7–13

unified machinery 102
universal quantifier 34, 98, 103, 118, 124, 148, 152, 153

vacuous quantification 34, 36–39
variable assignment 8, 104
variable binding approach 141
variable-free semantics 134
variables 29, 36, 37; binding 1, 27–38; case against 127–129; case for 122–127; in counterpart semantics 39–44; in epistemic contexts 44–54
variadic functions 70–72, 75, 82
Velupillai, V. 79
Vlach, F. 97, 124

waj (inferential non-future) 67
weakly intensional operator 5
Wehmeier, K. 2
well-formed expressions (wfe's) 7, 9, 10, 18, 20, 22
wfe's *see* well-formed expressions (wfe's)
Williams, V. 80
Wiltschko, M. 64, 68, 75, 76
Wittgenstein, L. 17n19
world reference 100–106

Yalcin, S. 32, 50
Yanovich, I. 2n1, 147, 157, 159, 161

Zimmermann, T. E. 2, 3, 164
Zwarts, J. 71

9781032580968